I0037892

Toxic Gas Sensors and Biosensors

Edited by

Inamuddin[1], Tauseef Ahmad Rangreez[2], Mohd Imran Ahamed[3] and Rajender Boddula[4]

[1]Department of Applied Chemistry, Zakir Husain College of Engineering and Technology, Faculty of Engineering and Technology, Aligarh Muslim University, Aligarh-202 002, India

[2]Department of Chemistry, National Institute of Technology, Srinagar, Jammu and Kashmir 190006, India

[3]Department of Chemistry, Faculty of Science, Aligarh Muslim University, Aligarh 202 002, India

[4]CAS Key Laboratory of Nanosystem and Hierarchical Fabrication, National Center for Nanoscience and Technology, Beijing 100190, PR China

Copyright © 2021 by the authors

Published by **Materials Research Forum LLC**
Millersville, PA 17551, USA

All rights reserved. No part of the contents of this book may be reproduced or transmitted in any form or by any means without the written permission of the publisher.

Published as part of the book series
Materials Research Foundations
Volume 92 (2021)
ISSN 2471-8890 (Print)
ISSN 2471-8904 (Online)

Print ISBN 978-1-64490-116-8
eBook ISBN 978-1-64490-117-5

This book contains information obtained from authentic and highly regarded sources. Reasonable efforts have been made to publish reliable data and information, but the author and publisher cannot assume responsibility for the validity of all materials or the consequences of their use. The authors and publishers have attempted to trace the copyright holders of all material reproduced in this publication and apologize to copyright holders if permission to publish in this form has not been obtained. If any copyright material has not been acknowledged please write and let us know so we may rectify this in any future reprints.

Distributed worldwide by

Materials Research Forum LLC
105 Springdale Lane
Millersville, PA 17551
USA
https://www.mrforum.com

Manufactured in the United States of America
10 9 8 7 6 5 4 3 2 1

Table of Contents

Preface

Many industrial and natural processes involve toxic gases, inflicting air pollution, living beings' health, and environmental crises became the most serious issues around the globe. It's vital that precise and speedy detection, forewarning, and scrutinizing of toxic gases is obligatory for environmental protection, air quality assurance, industrial and automotive sectors, etc. Toxic gas sensors and biosensors act as crucial role in numerous features of technology, industry, environment, safety, and a healthy lifestyle. Therefore, awareness and an in-depth understanding of sensing technologies to target toxic gases with increased sensitivity and excellent selectivity are mandatory.

This book highlights the recent advances and developments in toxic gas sensors and biosensors. Following a brief overview of the current status of toxic sensing technologies, the book scrutinizes the different toxic gases-based sensors, catalytic nanomaterials from 1D to 3D, principles, mechanism actions, analytical tools, various concepts, challenges, future scope, and directions are addresses. It also includes novel materials-based sensors, flexible sensors, scale-up and commercialization in the market, applications in healthcare devices, medical, environmental monitoring, consumer, industrial, cyber-physical system, and among others.

This book is envisioned for senior graduate students, faculty, researchers, and engineering professionals involved in research and development of sensing technologies.

Key features:

- Provides a comprehensive picture of gas sensors
- Featuring key contributions from well-established experts
- Explores recent developments and cutting-edge technology for gas sensors

Summary

Chapter 1 focuses on the detection and sensing techniques of NO_2 gas. Many adsorbents for example oxides of metals, perovskites, zeolites, spinels, and carbon-containing materials are discussed. A variety of gas sensors including electrochemical, catalytic combustion, semiconductor, and solid electrolyte gas sensors, are also reviewed in this chapter.

Chapter 2 discusses basic information of sensors, suitable conditions for efficient performance, and their development over time. Detection of various gases via graphene-based gas sensors are discussed in detail. This chapter specifically focuses on the analytical performance of CNT/Ni nanocomposites over glucose biosensors to acknowledge the feasibility of carbon materials beyond graphene. Additionally, the classification of CNTs, their fabrication mechanisms, and various fields of application are also highlighted in this literature.

Chapter 3 discusses the uses of 2D nanomaterials e.g. graphene, black phosphorus, transition metal dichalcogenides, etc. for gas and biosensing. Especially, their applications in the SPR sensor for toxic gas and biosensing are discussed in detail. Finally, we proposed a P3OT thin film-based SPR sensor for sensing of different concentrations of NO_2.

Chapter 4 elucidates the myriad preparation techniques on both the single and multi-layered stacked MXenes as well as their distinctive structural and electron characteristics. Selective membrane separation, adsorption, and photodegradation are given a focus for the MXenes in the field of gas removal. Besides small molecules and gases, MXenes as biosensors is also presented with wide potential applications in biomedical detection and environmental analysis.

Chapter 5 discusses various methods of hydrazine determination. Among these methods, the electrochemical method shows high sensitivity and selectivity towards the detection of hydrazine. Hydrazine is used in many industries such as agriculture, power generation, pharmaceutical, aerospace, and chemical industries. Hydrazine is fatal to both environmental and human life.

Chapter 6 describes a detailed literature review of the design, fabrication, and development of electrodes using graphene and heteroatom incorporated graphene-based nanocomposites for glucose sensing. It discusses the comparative study of sensitivity, stability, reproducibility, and low detection limit of functionalized and non-functionalized graphene derivatives for nonenzymatic electrochemical detection of glucose.

Chapter 7 discusses the general characteristics of biological sensors and the role of those devices in detecting toxins in the environment. Additionally, the future trends for developing sensors by utilizing nanotechnology may contribute to recognizing different poisons with great accuracy surpasses conventional patterns.

Chapter 8 is focused on the basic theory of chemical sensing and optical sensor arrays. The readers will find current literature survey on preparation methods using thin-films, polymers, monoliths, and pellets for sensing arrays. Furthermore, selected applications in environmental, pharmaceutical, medical, and food sectors are reviewed.

Toxic Gas Sensors and Biosensors   Materials Research Forum LLC
Materials Research Foundations **92** (2021) 1-38   https://doi.org/10.21741/9781644901175-1

Chapter 1

Nitrogen Dioxide Sensing Technologies

Shabbir Hussain[1], Khalida Nazir[2], Ata-ur-Rehman[3], Syed Mustansar Abbas[4]*

[1]Department of Chemistry, Lahore Garrison University, Lahore, Pakistan

[2]University of Sargodha, Women Campus, Faisalabad, Pakistan

[3]Department of Chemistry, Quaid-e-Azam University, Islamabad, Pakistan

[4]Nanoscience and Technology Department, National Centre for Physics, Islamabad, Pakistan

* qau_abbas@yahoo.com

Abstract

The harmful impacts of nitrogen dioxide (NO_2) include acid rain, respiratory diseases, allergy and photochemical smog. These also causes throat, eye and nose problems, cough, nausea and tiredness in extremely low concentrations (<10 ppm). So, detection and the sensing of NO_2 gas is considered as one of the most important detecting techniques. Numerous electronic sensors, semiconductor nanomaterials, carbon nanotubes, graphene, activated carbon and mixed metal oxides have been investigated in order to detect and sense NO_2. Several varieties of gas sensors including electrochemical, catalytic combustion, semiconductor and solid electrolyte gas sensors, have been industrialized.

Keywords

Nitrogen Dioxide, Sensors, Wet Oxidation, Carbon Nanotubes, Graphene

Contents

1.   Introduction

One of the most serious environmental issues is the emission of harmful substances like oxides of nitrogen into the atmosphere. With the greater increase in the population of the world, the consumption of fuel has also been increased at the same speed. However, the burning of fossil fuels emits many toxic gases including oxides of carbon, sulfur and nitrogen which affect the air quality especially in the industrialized areas [1, 2]. Currently, the release of nitrogen oxides (NOx where X = 1 or 2) from the burning of oil, gas and wood in industrial units and diesel engines of vehicles has resulted in serious environmental issues [3, 4]. The other source of nitrogen oxides is nitrogen-fixing bacteria [5]. The harmful impacts of nitrogen oxides (NO or NO$_2$) include acid rain, respiratory diseases, allergy [6-8] and photochemical smog [9]. The respiratory problems may include the increased risk of respiratory conditions, decreased lung function caused by long term exposure, respiratory inflammation of the airways, decreased absorption of oxygen molecules on red blood corpuscles and lung irritations [10, 11]. Lung damage has been reported by its long-term exposure in animals [12,13]. It may also develop conditions like bronchitis and emphysema [14].

Nitric oxide (NO) combines with ozone and forms nitrogen dioxide (NO$_2$) with the release of oxygen (Eq. 1) [15]:

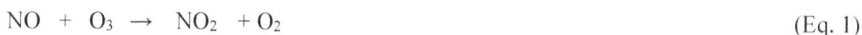

$$NO + O_3 \rightarrow NO_2 + O_2 \qquad\qquad\qquad\qquad (Eq.\ 1)$$

The ground level ozone formed in this way causes serious damage to the environment and its associated lives. It is worth mentioning that ozone (O_3) is naturally formed in the upper layer of the atmosphere and protects the earth from the ultraviolet rays of the sun. However, the formation of unnatural ozone at ground level poses serious threats to human beings [16].

NO from engines of vehicles is one of the most prevalent NOx species. It is oxidized in the atmosphere in order to be converted into NO_2 and also reacts with most hydrocarbons to form ozone; both are strong oxidants and toxic. Thus NOx is a criteria pollutant [17]. NO_2 is a reddish brown gas and has a pungent smell, is highly volatile and dangerous to humans and the ecosystem [7,16,18-24]. NO is unstable and rapidly forms NO_2.

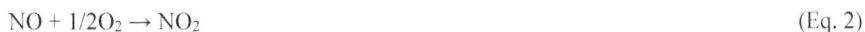

$$NO + 1/2O_2 \rightarrow NO_2 \qquad\qquad\qquad\qquad\qquad (Eq. 2)$$

A smaller concentration of NO_2 is not dangerous for human beings, but when its concentration is increased to more than $200\mu gm^{-3}$ it becomes harmful [7]. NO_2 gas may easily react with other pollutants under photochemical reactions [25-28] and is also produced as a secondary pollutant in the atmosphere [29-31]. NO_2 also causes throat, eye and nose problems, cough, nausea and tiredness even when its concentration is extremely low (<10 ppm) [11,32,33]. Due to these reasons, the sensing of NO_2 gas is considered as one of the most important detecting techniques.

The oxides of nitrogen which are primarily discharged from combustion processes and car exhausts are considered as important pollutants; their higher concentration may result in the production of ground level ozone [34]. To eliminate or minimize the emission of nitrogen oxides special precautions should be taken. Numerous techniques including selective non-catalytic reduction (SNCR), selective catalytic reduction (SCR) and NOx storage reduction (NSR) are applied [35]. Currently every state of the world is making rules and laws to reduce the air pollution by adopting various strategies and procedures [36,37]. The demand of sensors for detection of air pollutants (e.g., NOx gases) has been increased dramatically especially with respect to global issues of environment and earth atmosphere. They are especially required in order to monitor and control the combustion systems of automobiles and industry [15].

Current studies were performed to overview various sensing techniques and methods used for the detection and sensing of nitrogen oxides. The study covers the use of carbon nanotubes (CNTs), graphene, activated carbon and mixed metal oxides for the purpose of NO_2 sensing.

2.   Sensing of NO_2

Among all the gases, the sensing and measurement of NO_2 is of great interest because nitrogen oxides produce huge environmental issues. The amount of NO_2 in the environment can be checked by different kind of gas sensors. It was reported that diesel engine emits a greater amount of NO_2 as compared to the gasoline engine. So NO_2 sensors are very important components in the diesel engines [3, 38-40]. These sensors may be classified into resistive sensors and electrochemical sensors [41-43]. The former is based on semiconductor while the later works on the principle of solid electrolysis. The versatile properties of semiconductor nanomaterials and simple adsorption principle for detecting a gas have attained very much attention [44-46]. It is fundamental for the researchers to plan and manufacture the gas sensors which are of minimal effort, precise, specific and delicate particularly for detecting of very low concentration of gases. Many adsorbents for example oxides of metals, perovskites, zeolites, spinels and carbon-containing materials are used to remove oxides of nitrogen from gaseous materials [47].

Several varieties of gas sensors including electrochemical, catalytic combustion, semiconductor and solid electrolyte gas sensors, have been industrialized. Out of these, semiconductors types of sensors are extensively used nowadays. These are neither too costly nor they require heavy technology for manufacturing [48,49].

There are many sensing technologies for the detection of gases emitted from the exhaust of vehicles. Major gases emitted from the exhaust are oxides of carbon (COx), oxides of sulphur (SOx), oxides of nitrogen (NOx) and hydrocarbons (HCs). The sensing of these gases isalso interconnected with the efficiency of the engine. For the detection and monitoring of the very lower concentration of pollutants, sensors should be highly sensitive, selective and accurate [50].

The detection of NO during high-temperature combustion [51-55] is an important area of research in the clinical analyses [56]. Sensing/detection techniques for NO commonly use the principles of electrochemistry [57], chemiluminescence [58], chromatography [59], mass spectrometry [60] and optical spectroscopy [61]. NO_2 sensor detection range is 0.5-5.0 ppm in cabins of the automobile. So, there is a need to improve the detection limit upto this concentration [1, 9].

Both NO and NO_2 have high oxidizing power so they react with metal oxides lying on the surface to produce oxygen. The reaction happening at the metal oxide's surface can be represented as follows [9]:

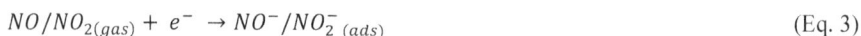

$$NO/NO_{2(gas)} + e^- \rightarrow NO^-/NO_2^-{}_{(ads)} \qquad\qquad (Eq. 3)$$

$$NO/NO_{2\,(gas)} + O_{2\,(ads)}^- \rightarrow NO^-/NO_{2\,(ads)}^- + O_{2\,(gas)} \qquad \text{(Eq. 4)}$$

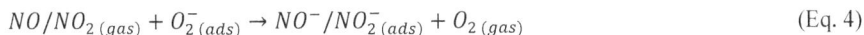

There are different sensing devices for NO_2. Some of them are explained below:

3. Electronic sensor

It has been well established that adsorbing and desorbing of gas molecules on the surface of metal/metallic oxide can change the electrical properties (resistivity, conductivity) of a material. For example, the electrical properties of ZnO are changed by adsorbing and desorbing the gas molecules. Since then, electronic semiconductors are extensively used in the sensing of polluted gases in the environment. There are so many advantages of these nanostructured particles but the most attractive advantage is the low cost of these particles and easy manufacturing. In addition to these, they are highly sensitive in their working [62, 63].

3.1 Working principle of the electronic sensor (semiconductor)

The energy band or energy gap theory states that there are two energy bands (conduction and valence bands) in the metallic conductors. These bands are separated from one another with respect to energy. The valence band has low while conduction band has high energy. So electrons in the valence shell need a sufficient amount of energy to occupy the conduction band. When electrons acquire that specific amount of energy they occupy the conduction band and conduct the electricity [64,65]. For many years these semiconductors were extensively employed for gas sensing due to changing of their electrical properties. As NO_2 gas is adsorbed and desorbed from the surface of semiconductors, it changes its conductivity which is measured [66, 67].

4. Carbon nanostructures and nanohydrates in NO_2 sensing technology

It has been well understood that CNTs are highly efficient for NO_2 sensing purposes. They can sense very small changes in NO_2 concentration and demonstrate very fast response [68-72]. Single-walled carbon nanotubes have low detection limit up to 25 ppb at 400 °C whereas while double-walled carbon nanotubes have detection limit as low as 100 ppb [73]. CNTs have played a key role in reducing human health problems and environmental exposure [74]. Tiny molecular films with increased surface area, act as sensors with a greater response, high sensitivity and good recovery time as compared to conventional sensors. Their sensing performance has an important correlation with microstructure, electrical properties and morphology of nanomaterials [1]. 1D and 2D CNTs along with graphene demonstrate extra-ordinary physiochemical properties because of their peculiar atomic and molecular arrangements so they are sensitive towards small changes in the neighboring gas atmosphere because their resistance and conductance changes largely with respect to adsorption of NO_2 gas molecules [73]. These gas sensors have core sensor-array and come together with both functional carbon nanostructures as well as with commercial gas sensors [75]. Solid-state gas sensors are very cheaper and economical for the quality control of the air. Calibration and validation are very compulsory for low-cost gas sensors for future prospects [76]. There are reports regarding the monitoring of landfill gas at 150 °C by miniature CNT-based gas sensor [77]. Surface acoustic wave (SAW) equipment (such as nanocomposites) having passive, miniaturized and gas sensing properties are suitable for gas-sensing technologies [78]. Nanostructured thin films of cross-linked polyaniline are fabricated with non-hydrated and hydrated WO_3 nanoparticles, for their uses as gas sensing tools [79]. The nanohybrids of $In(OH)_3/Co(OH)_2/rGO$ can be prepared by simple reflux method. The surface area of nanohybrids is increased by adding indium to $Co(OH)_2$ on rGO; in this way, the gas sensing response to NO_2 becomes higher than that of $Co(OH)_2/rGO$. Excellent gas-sensing properties are exhibited with amass ratio of 5:1 in these nanohybrids. Thus, at room temperature, high sensitivity of 75.32% and low detection limit (0.97 ppm) of NO_2 can be achieved. This sensor also exhibits excellent NO_2 selectivity even when other gases (e.g., hydrogen, carbon monoxide, acetylene, ammonia) are present. By indium doping, the charge carrier concentration of cobalt hydroxide is increased which promotes the transfer of electron when the gases are interacted [80, 81]. A large variety of carbon nanomaterials were found effective for sensing of NO_2 [73,82]. Due to their important electrical and physical properties, carbon nanomaterials such as graphene [83, 84], CNTs [85-87], carbon fibre (CF) [88, 89], fullerene [90] and carbon black (CB) [91] have got a considerable attention as gas sensing materials. The important characteristics such as high rigidity, high heat or electrical conductivities and high surface-to-volume proportions make the carbon nanomaterials as ideal NO_2 sensors [83].

4.1 Carbon nanotubes

CNTs find applications as sensors due to their amazing properties, for example, quality and soundness [84,92], larger surface (>1500 m^2 g^{-1}) and high electrical conductivity. These sensors are generally superior over traditional sensors (based on metal oxide) due to low energy utilization, good activity at room temperature and also to some degree higher gas reaction [9].

Due to their high thermal conductivity and pi-electronic distribution, single-layer and multi-layer CNTs demonstrate good ability to sense the NO_2 from the environment [93, 94]. The Pi electronic distribution provides the excess of electrons so that the ability of

NO_2 to accept the electron is enhanced and NO_2acts as an oxidizing agent. Charge transfer takes place between NO_2 and CNTs. The amount of adsorbed species can be calculated by measuring the conductance because there is a direct relationship between the conductance and the number of species which are adsorbed [95].

Doping of semi-metals and non-metals (like boron and nitrogen) on carbon multi-layers increase the sensing properties by changing the energy gap between HOMO and LUMO molecular orbitals. Dopants modify the electrical and physical properties of template molecules so a very low concentration of NO_2 (50ppb) can easily be detected at room temperature. The increase in binding energies probably enhances the charge transfer between N-doped multilayer CNTs and NO_2 gas molecules [96]. The presence of heteroatoms such as nitrogen, oxygen or sulphur makes the surface of CNTs very active for the incoming group [97].

NO_2 sensing studies by fabrication of CNTs are currently under common investigations. CNTs are considered a valuable sensing substance due to their very important properties i.e., wide surface area, high adsorption ability, high aspect ratio, good mechanical strength and high electrical conductivity. Excellent NO_2 detecting is reported with single-wall carbon nanotubes [98]. The response of Al/CNT sensors is enhanced suddenly upon the NO_2 exposure [99]. It is considered that the Schottky barrier may be changed at the Al/CNT interface for the purpose of NO_2adsorption of molecules. The Al/CNT sensor has a faster Schottky response as compared to that of CNTs [100].

4.2   Graphene

Graphene contains honeycomb crystal lattice having a two-dimensional single layer of sp^2-hybridized carbon atoms [101]. Every atom of single-layered graphene sensors is present at the surface so sensitivity is displayed down to the single molecular level. Their sensitivity was attributed to their metallic conductivity, even when very few charge carriers are present. The gas sensors of multi- or monolayered graphene may be grown in various ways i.e., by reducing graphite oxide (RGO) etc. Monolayered graphene is more sensitive to NO_2 gas. It is well understood that the graphene-based sensors demonstrate improved sensing performance by patterning the graphene surface into nanostructured mesh or ribbons or its modification with metals and conducting polymers. By using the gating technique or the analysis of low-frequency noise spectra, sensitivity and selectivity can be further improved. Epitaxial graphene has been reported as a superior gas sensor as compared to the other graphene. Epitaxial graphene with single-layers has natural ability to receive the electron density from the SiC substrate so it is considered as a semiconductor of n-type. It has been well understood that n-kind graphene sensors exhibit better sensing performance as compared to the p-type [102]. A sensor can be assembled along with a computer interface and a sampling system into a portable unit in order to suitably monitor the NO_2 concentration [103].

The $Pd\text{-}SnO_2$-reduced graphene oxides are highly sensitive to NO_2 [8]. Table 1 [73] shows a comparison among different materials based on graphene devices for sensing and detecting the NO_2 gas. The varying sensitivity is due to the different materials used. Current operational mode devices made up of materials like reduced graphene oxides have the ability to sense NO_2 with LOD ~2 and response time >600. Conductivity operational mode device made up of materials like graphene flakes, exfoliated graphene nanostructure and graphene films can sense/detect the NO_2 gas up to a certain limit (0.06 and <0.01) and response time (~6, <12 and 50-100), respectively.

Table 1   *Sensing characteristics of graphene-based devices employed to detect NOx gases. Reproduced with permission [73] Copyright (2013) Elsevier*

Material	Device operational mode	Target analyte	Limit of detection (ppm)	Response time (s)	Recovery time (s)
Mechanically exfoliated graphene	Resistivity	NO_2	~0.001	~6	-
Abraded nanoscale graphene flakes	Electrical conductance	NO_2	0.06	–	–
Mechanical exfoliated graphene nanostructures	Conductance and surface work function	NO_2	0.06	<12	–
Graphene oxide reduced with hydrazine	Resistance	NO_2	<5	<600 at RT-150 °C	<600 at RT-150 °C
Low temperature thermally reduced graphene oxide	Current	NO_2	~2	>600 at RT	>600 at RT
Graphene films and ribbons via microwave PECVD	Resistance	NO_2	<100	>600 at RT-200 °C	–
Epitaxial grown uniform graphene films	Conductance	NO_2	<0.01	50–100 at RT-250 °C	–
Chemically exfoliated graphene from natural graphite	Resistance	NO_2	<0.35	<600 at RT	<600 at RT

Reduced graphene oxides provide better sensing for NO_2 even at low concentration [104]. Gallium nitride sensors also exhibit good sensing capacity for NO_2 as BGaN/GaN can detect NO_2 up to ppb level with quick response time [8]. There were investigations on the role of gas sensors based on graphene/metal-oxides for detection of CO, NH_3, NO_2 and also some volatile compounds of organic nature. The metal-oxide gas sensors have the ability to respond/sense well to the environmental concentrations of CO, NO, NO_2, O_3 [105].

4.3   Activated carbon

NO_2 can be removed by various techniques including selective catalytic reduction (SCR), selective non-catalytic reduction (SNCR) and NO_2 storage reduction (NSR) [106]. SCR and SNCR techniques are highly useful for successful elimination of nitrogen oxides. However, both SCR and SNCR suffer from the following limitations:

i. Costly

ii. Technical complexity

Therefore, separation is mostly favoured by adsorption. Zeolite, perovskites and carbonaceous things are tested for treatment of NO_2. The NO_2exhibits good adsorption coefficient when activated carbon (AC) is employed [107]. Appropriate sized mesoporous activated carbons are designed which have oxygen groups attached on the surface so that NO_2gets adsorbed on the surface of activated carbon (AC) and is separated [47]. On the AC surface, heteroatoms complexes are formed which enhance the adsorption potential [97]. NO_2 can be adsorbed on the AC surfacethrough reduction, chemisorptions or/and physisorption which results in the production of different surface complexes e.g., C-O, $C-ONO_2$, C-ONO, $C-NO_2$ etc. There were investigations on activated carbons with various porous amounts, textures and the nature of surface oxygen groups in order to determine the role of oxygen surface groups in NO_2 adsorption. Activated carbons can be prepared from different kinds of lignocellulosic precursors for elimination of NO_2 at the ambient temperature [108,109].

A fixed-bed flow reactor having 0.05 nm diameter was used for sensing of NO_2; the experimental set up [107]. Two nitrogen oxide analyzers were fitted at the outlet and inlet of the reactor using chemical luminescence in order to determine the concentrations of NO and NO_2. The column was filled with activated carbon (2g) in each test. Before the tests, however, it is necessary to condition the activated carbon samples for 15 min in the airstream at 23 °C and 50% relative humidity [110]. For the preparation of modified AC having various surface chemistry, different protocols can be applied as discussed below [111].

4.3.1 Wet oxidation

To 250 mL aqueous solution of 1 M nitric acid, chemically activated carbon (30 g) is mixed and refluxed for 8 hours. The obtained mixture is filtered and washed with distilled water to adjust the pH value of filtered water around 7.

4.3.2 NaOH neutralization

To the 1 M solution of sodium hydroxide, chemically activated carbon (50 g) is added followed by refluxing for three hours. After filtration, the mixture is washed with distilled water; it is finally kept at 105 °C overnight for the purpose of drying.

4.3.3 Thermal treatment

The activated carbon is left for 1 hour under the nitrogen flow of 10 NL h^{-1} at 900 °C; it is then cooled to the room temperature. This treatment removes the surface oxygenated functional groups.

Wood-based activated carbon BAX-1500 can be employed to adsorb NO_2 in dry or wet conditions in a glass column. NO_2 is absorbed and reduced on the surface of activated carbon. The sensing capacities of BAX-1500 wood-based activated carbon were observed to be 63.8 and 42.7 mg g^{-1}. The adsorption causes an increase in acidity of AC and reduces its textural properties [112]. The chemistry of surface has a very crucial role in adsorption/sensing of NO_2. After the adsorption of NO_2, BAX-1500 is converted into the powder form. Uptake/sensing of NO by activated carbon varies in different atmospheres. In the absence of oxygen, less amount of NO is adsorbed [113,114] because oxygen is necessary for maximizing the NO adsorption. Firstly, AC is exposed to oxygen by itself. Then adsorption of oxygen takes place. After adsorption, the weight of carbon is increased up to 0.5 wt.% and 50% of NO is adsorbed by the activated carbon. Same adsorption potential for NO is observed in CO_2/O_2 atmosphere as in O_2 environment [115].

NO_2 adsorption is considerably affected by the surface chemistry and porosity of activated carbons [116]. The accessibility of the molecules of NO_2 into micropores especially affects the adsorption capacity while the NO_2 adsorption/reaction has been mainly affected by the surface chemistry [116].

There are two structural features which decide about the basic character of activated carbon:

i. Presence of basic amines or basic surface functionalities (pyrone, ketone, chromene)

ii.   The presence of oxygen-free Lewis basic sites on graphene layers

The first procedure involves the heating under an inert atmosphere of He, N_2 andH_2 gases at high temperature to eliminate the acidic groups of surface selectively. Nitrogenation is involved in the second procedure; here the nitrogen atoms are incorporated on the surface of activated carbon. The basicity of activated carbon is enhanced due to the creation of amine bases by this procedure. The long term exposure and higher temperatures convert the nitrogen groups into imide and imine and then into pyridine and nitrile [117-119].

Activated carbon can be applied as an efficient adsorbent in order to lower the NO_2 emissions in gaseous run-offs [120,121]. It can be prepared on commercial level by chemical/physical activation of different chemicals e.g., mineral coals or lignocellulosic biomasses [122]. It was observed that adsorption, as well as reduction of NO_2, can take place on the surface of activated carbon. The same kind of behaviour was reported in2004 during studies on the interaction between NO_2 and carbon black [112]. The adsorption capacities of activated carbon were observed to be 63.8 and 42.7 mg g^{-1} in a wet and dry environment, respectively. Here the acidic properties of the activated carbon were increased by NO_2 adsorption; it was due to the formation of nitric acid and the surface oxidation when NO_2has interacted with hydroxyl groups. The textural properties i.e., specific surface area and micropores volume were also lowered to 10–15% by surface oxidation during NO_2 exposure. So it was clarified that both the surface chemistry and textural properties play vital a role in NO_2 adsorption/reaction with activated carbons. The preparation of activated carbon from polish bituminous coal may involve the pyrolysis of the coal precursor at 500–600–700 °C and then activating at700 °C by KOH treatment [123]. During the adsorption, the NO_2 is reduced into NO. The adsorption capacities for pyrolyzed activated carbon were enhanced from 0.5 and 0.7 mg g^{-1} at 700 °C to 43.5 and 25.0 mg g^{-1}in wet and dry conditions, respectively. It was found that the adsorption capacity was increased by high microporous volume achieved during KOH activation and also the well-developed porous structure [123].

The role of surface chemistry of activated carbon was clarified during experiments on CO_2 physical activation of sawdust pellets at different activation times (30–120 min) and temperatures (600–800 °C). When the activation time period was enhanced at 800 °C from 30 to 90 min, the adsorption was enhanced to 45.3 mg g^{-1} from 18.6 mg g^{-1} [122, 124]. Similar results were obtained during experiments on other activated carbons i.e., from lignocellulosic biomasses e.g., plum stone and walnut shells using KOH chemical activation [125, 126]. In the same context, there were studies on commercial activated carbons prepared from date pits-modified activated carbons by heat treatment and chemical oxidation [116]. It was shown that the crucial step in NO_2 adsorption is the reduction of NO_2 into NO [112, 116]. However, the textural properties also have a considerable role in this regard [112].

As far as the role of surface chemistry is concerned, there is no any clear correlation between the adsorption capacity and the number of surface oxygen groups. The reason is that such kind of correlations demand the activated carbon (for experiments) having analogous textural characteristics and only the surface chemistry modified. Also, the role of basic and acidic surface groups is not clearly differentiated [122,124,125,127,128] because both the groups undergo changes when NO_2 interacts with the carbon surface. However, the NO_2 adsorption may be inhibited when carboxylic groups are present in high concentration on the surface of activated carbon.

5. Metal oxides in NO_2 sensing

ZnO, SnO_2, In_2O_3, Nb_2O_5 and WO_3, are metal oxides which may be employed as gas sensors. Here the detection/sensing of a target gas is associated with a redox reaction which takes place between the metal oxide surface and target gas [129]. ZnO, TiO_2 and WO_3 are metal oxide semiconductors of n-type and the efficiency of these sensors is greatly affected by the composition of chemicals, microstructure, thickness and crystallographic features such as phase constitution, density and thickness [130-132]. The most active part is a thick and porous layer of the n-type. Increase of surface and decrease of the layer thickness to a minimum value (100 nm) will affect the sensing response, time for recovery and response time [133].

The material on which sensor is based can react with a target gas and modify its properties like electrode potential, dielectric constant, mass, work function or it may emit heat or light. The semiconductor sensors are of two types, oxides or non-oxide. The oxide sensors are able to suffer the harsh conditions and are temperature resistant so they are suitable to be used as a transducer as well as receptors. While, it is necessary to cover the non-oxides sensors with an insulating layer which acts as a protective layer [15]. The semiconductors which have wide gap are useful because of their variety of applications. The high sensitivity of the semiconducting metal-oxides, as well as high compatibility and low cost, makes them more attractive [134,135].

CuO, SnO_2, In_2O_3, WO_3, TeO_2, Fe_2O_3, MoO_3, ZnO, CdO and TiO_2 nanostructures with different dimensions and sensor configuration are prepared to date. Morphology and surface are of metal oxide play a decisive role in their performance in gas sensing [136]. Metal oxides can be prepared to have different morphology of surface and different configuration such as thick films, thin films, one-dimensional nanostructure and single-crystal depending on the application of interest [137]. However, at room temperature, the humidity may affect metal oxide gas sensors [5]. Excellent chemical and thermal stabilities in different conditions of operating and high surface to volume ratio make these nanostructures well suited for applications [138-140]. Fabrication of one-dimensional nanocrystals has been of major interest in nanotechnology and nanoscience. There are different methods for preparing 1-D metal oxides [141]. These methods are ultrasonic irradiation [142], hydrothermal [143], electrospinning [144], sol-gel [145], anodization [146], molten-salt [147], solid-state chemical reaction[148], carbothermal reduction [149], chemical vapor deposition [150], nanocarving [151], molecular beam epitaxy [152], aerosol [153], RF sputtering [154], vapor-phase transport [155], thermal evaporation [156], dry plasma etching and UV lithography [157].

Indium oxide is considered as a valuable material due to its high electric conductance, and solid association with certain gaseous atoms. The gas sensors broadly employ In_2O_3 because of its appropriate potential to detect NO_2 [15].

Target particles of NO_2 gas give direct absorption onto $Co(OH)_2/rGO$ to create NO_3 by reacting with O. The NO gas atoms are adsorbed similarly and respond with O to produce NO_2 [158]. The sensing mechanism has been shown in Fig. 1 [158] while Fig. 2 [158] demonstrates the sensitivity of pure $Co(OH)_2 Vs Co(OH)_2/rGO$ for NO_2.

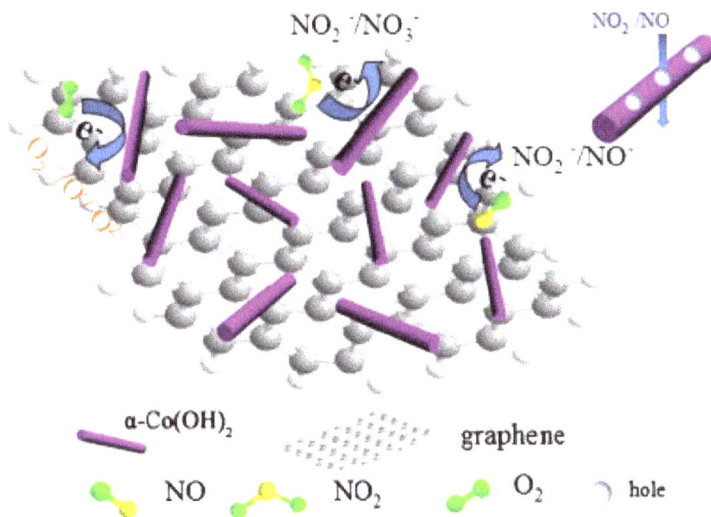

Fig. 1 *Mechanism of a sensor for $Co(OH)_2/rGO$. Reproduced with permission [158] Copyright (2014) Elsevier*

Materials Research Forum LLC
https://doi.org/10.21741/9781644901175-1

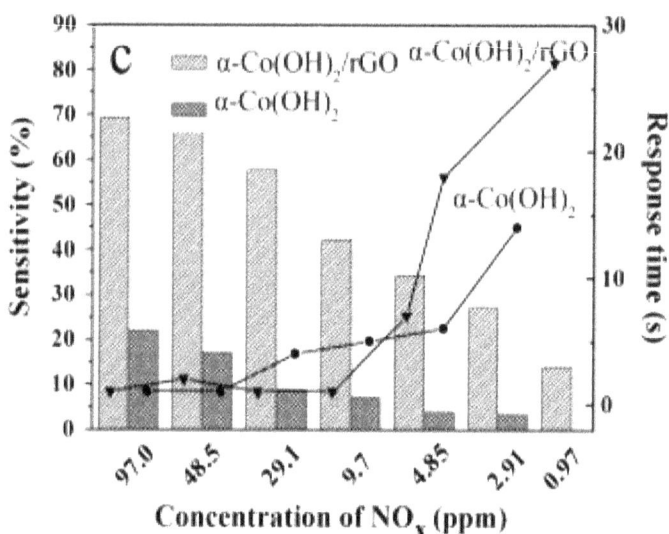

Fig. 2 *Sensitivity of Co(OH)₂/rGO and pure Co(OH)₂ Vs concentration of NOx.*
Reproduced with permission [158] Copyright (2014) Elsevier

A sintered film of WO_3 is highly sensitive to NO_2 as well as NO gas [41]. The maximum sensitivity was observed at 200 °C for 200 ppm of NO gas. This sensor exhibited an outstanding response for NO_2 gas sensing at 300 °C. The sensor has displayed a particularly low response for other gases like CH_4, hydrogen and CO in the fabric; it demonstrates that these sensors have the ability to differentiate between various gases. Several metal oxides including CuO, In_2O_3, WO_3, ZnO and SnO were investigated in order to monitor the presence of some gases specially from an exhaust [159]. These materials displayed different responses to the presence of NO_2 and NO; the responses were markedly changed with the change of temperature. At 200 °C, ZnO exhibited a really poor response; the CuO displayed no response whereas the In_2O_3 showed a very superior response. However, at 400 °C, the response was changed; most significant response to the presence of the gas was displayed by ZnO while CuO still exhibited no response. The response time at 200 °C lies in the range of 10–30 s [159].

The threshold limit value of NO_2 is 25 ppm so detection of its low concentration (10–200 ppm) is extremely important for the safety of the lives of human beings [160, 161]. Within the past, chemiresistive NO_2 sensors were developed. Thick/thin films of metal-oxides like NiO [162], WO_3 [163], In_2O_3 [27] and SnO_2 [164] were used. However, these metal oxides demonstrate inherent drawbacks of the requirement of a high operating temperature (4300 °C), limited detection range, long reaction time and poor selectivity. Low-cost chemical gas sensors can be obtained by fabricating nanostructured metal oxides having semiconducting properties and possess excellent sensitivity to harmful gases. In such materials, extremely small grain size results in a really large area of metal oxide for its contact with a surrounding gas; in this way good sensitivity to a large varieties of gases becomes possible [165,166]. The production of nanoscale metal oxides having reproducible properties (grain size, thickness, porosity, structure) plays the main role in semiconductor gas sensor [167]. α-Fe_2O_3 is an environmentally friendly device which possess excellent potential to detect/sense harmful, explosive, combustible and toxic gases in numerous applications (whether industrial or domestic) [168, 169]. However, the form and size of α-Fe_2O_3 majorly affect the gas sensing properties [170-172].

$NaNO_2$ gate and WO_3 gate devices use a MISFET transducer to sense NO_2 gas [15]. Organic semiconductors can detect a low level of NO_2. WO_3 nanowires have an excellent detection limit of 5ppm NO_2. Likewise, SnO_2 hollow spheres and nanowires can detect up to sub-ppm concentrations [8, 9]. Bi_2O_3 which is p-type semi-conductor possesses an excellent detection limit for NO_2 [9, 173]. Dopants increase the gas sensing response e.g., MoO_3 doped SnO_2 exhibits 3.6-500 ppm response to NO_2 [9]. Pd-doped TiO_2 at 180 °C has a response to 2.1 ppm NO_2 [9].

5.1   Mixed metal oxide in NO_2 sensing

Recently, many mixed metal oxide gas sensors have been prepared with various technologies and materials in order to detect various gas components. The mixed metal oxides based on various materials support the human's life in different ways. The metal oxides can sense the NO_2 gases and it is understood that a mixture of two metal oxides can change the electrical properties of sensors and their sensing performance. Nowadays, the scientific community focuses on the investigation of a different combination of metal oxides as gas sensors to achieve excellent selectivity and good response time for numerous gases [64, 174-181]. A nitrogen oxide based sensor is usually a high-temperature device. These sensors are made to detect NO_2 in combustion environments such as a truck tailpipe, automobile or smokestack [182]. Due to the wonderful sensitivity to gases, metal-oxide semiconductors are extensively applied for gas sensing purposes [183, 184]. Good chemical stability and excellent response towards NO_2 are associated by mixing together SnO_2 and WO_3. It was known that the response of ZnO-SnO_2 core-shell nanowires for NO_2 gas can be enhanced up to 33 times at 200 °C as compared to pure

ZnO nanowires; the threshold detection is lowered to to15 ppb at 200 °C and 138 ppb at 300 °C. The response of the composite of SnO_2 and WO_3 is much better as compared to pristine WO_3 and SnO_2 at 200 °C [179, 185-188]. In solid state devices mostly conductometric chemical sensors with semiconducting layers of metal oxides are used; they have low power consuming and are low priced. In the atmosphere, the reaction between gases and semiconductor rely on the sensing characteristics of a material. The presence of oxidizing/reducing gases strongly influences the resistance of semiconducting thin films [189].

For the detection of NO_2 gas, low cost, highly sensitive and least maintenance sensors are required [190,191]. The tungsten oxide, zinc oxide and tin oxide mixed metal-semiconductor is very responsive to NO_2 gas and chemically stable [192,193]. The sensitivity of SnO_2-ZnO semiconductor is 33 times superior as compared to the pure ZnO for NO_2 gas at 200 °C and threshold detection limit is reached to 138 ppb at 300 °C and 15 ppb at 200 °C [179]. The WO_3 and SnO have greater sensitivity than pure WO_3 and SnO at 200 °C. The nanofibers of ZnO and SnO_2 can be made by using the electrospinning method and laser pulse deposit [194]. This nanofiber is highly sensitive and can detect very low concentration up to 400 ppb as compared to pristine ZnO nanofiber [195].

Almost all the semiconductors work on the same principle. When the gas for which sensor is designed reaches on its surface its resistance either decreases or increases because the gas covers its surface due to adsorption. Intact, the electron transfer occurs between semiconductor and gas absorbed on it which causes an alteration in the charge transfer within the sensor. For example, when an electron moves from the sensor to the NO_2 gas then it changes to anionic form at semiconductor. In case of the n-type sensor, the conductance decreases due to lowering of the number of electrons on its surface and vice versa for a semiconductor of p-type. This is the simple principle to detect the gas by measuring the alteration in the conductance which is made due to the measurement of resistance of two electrodes by comparing with some analyzer. Such sensors are very useful, cheaper, small in size, easy to integrate into electronic devices and require minimum maintenance [185].

Metal-insulator-semiconductor field-effect transistors can be employed in which metal oxides like $NaNO_3$ and WO_3 act as receptors or sensors and signals are produced in the form of threshold voltage shift and are used to detect NO_2 gas. Examples include sodium nitrate gate field effect transistor and tungsten trioxide gate field effect transistor [196]. Hybrid materials of metal oxides with other semiconductors may offer a greater possibility of detection for future demands [9]. Nanomaterials in the hybrid form are surprisingly sensitive to NO_2. Table 2 [197] demonstrates the average response and standard deviation for NO_2 by different sensors [197]. The data have been graphically explained in Fig. 3 [197] and Fig.4 [197]. The Table 1 [73] shows that SnO_2 with CNTs produce the best response when present in medium or low concentrations.

Table2   *Average response and standard deviation for NO_2 by different sensors. Sensor operation occursat Normal temperature and at 150°C. Reproduced with permission [197] Copyright (2010) Elsevier*

Temperature	NO_2 concentration (ppb)	CNT/SnO₂ high	CNT/SnO₂medium	CNT/SnO₂ low
Normal	50	<0.01	2.77 (0.0003)	0.68 (0.0027)
	100	0.039 (0.00004)	4.59 (0.0046)	1.24 (0.0050)
150°C	50	<0.01	0.06 (0.0001)	0.63 (0.0008)
	100	<0.01	0.06 (0.0001)	1.02 (0.0013)

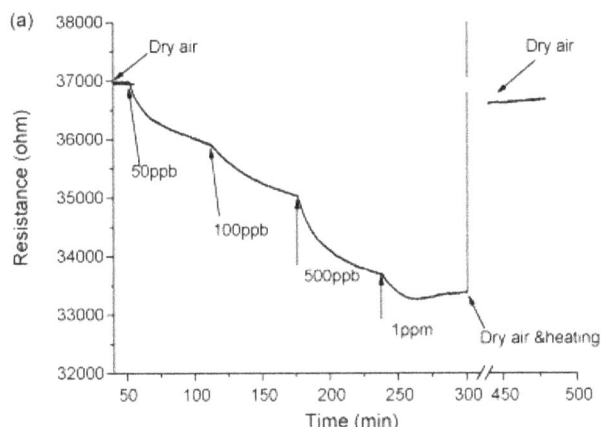

Fig. 3 *At room temperature, the resistance of the sensor becomes low with increasing NO_2 concentration. Reproduced with permission [197] Copyright (2010) Elsevier*

Polythiophene/SnO_2 composites are utilized for detecting oxides of nitrogen at various temperatures. The nanocomposites of PTP/SnO_2 can detect oxides of nitrogen even at a lower concentration of 10 ppm. Their sensitivity is graphically represented in Fig. 5 [198] and Fig.6 [198].

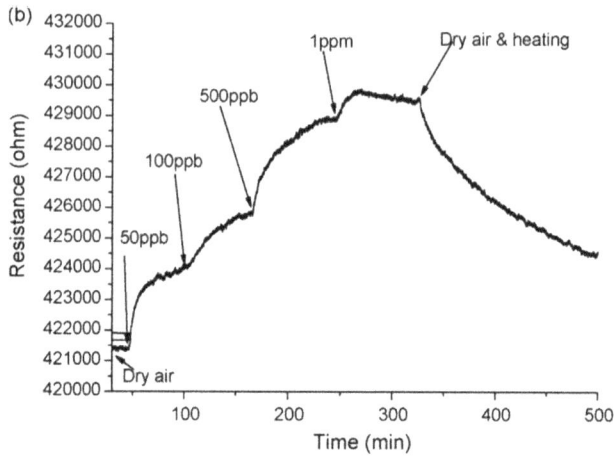

Fig. 4 *At 150°C, the resistance of the sensor becomes high with increasing NO₂ concentration. Reproduced with permission [197] Copyright (2010) Elsevier*

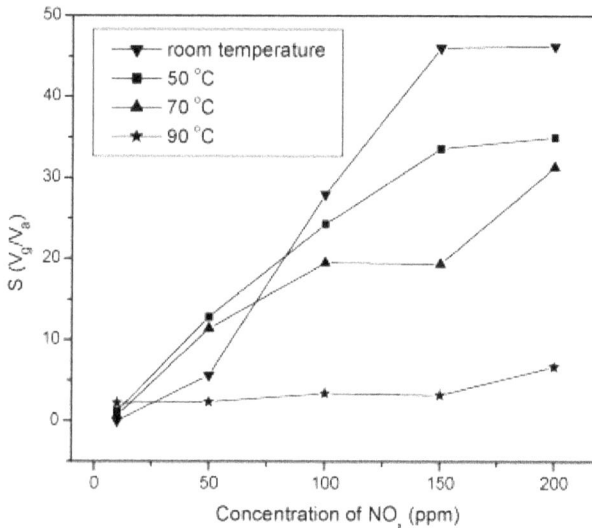

Fig. 5 *5% PTP/SnO₂ sensitivity for the concentration of NO and NO₂. Reproduced with permission [198] Copyright (2008) AIP Publishing*

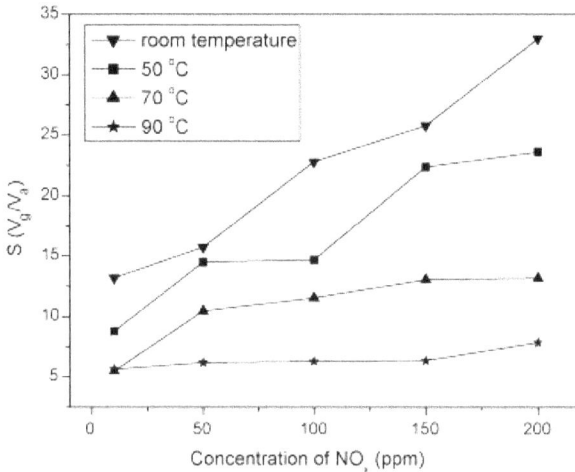

Fig. 6 *10% PTP/SnO₂ sensitivity for the concentration of NO and NO₂. Reproduced with permission [198] Copyright (2008) Elsevier*

Conclusions

Nitrogen dioxide (NO₂) pose serious threat for human being in the form of acid rain, photochemical smog, respiratory diseases, allergy, throat, eye, nose problems, cough, nausea and tiredness. Numerous materials including semiconductors, carbon nanotubes, graphene, activated carbon and mixed metal oxides are used to detect and senseNO₂. The response depends upon temperature, surrounding atmosphere, ease of transfer of valence electrons into the conduction band and surface chemistry of a sensor. There is also the choice of integrating a sensor with the computer interface and a sampling system into a portable unit for gas monitoring. Many gas sensors possess an excellent ability to differentiate between various gases. A sintered film of WO₃ displays an outstanding response for NO₂ gas at 300 °C but low response for other gases like CH₄, H₂ and CO. The In(OH)₃/Co(OH)₂/rGO nanocomposite have the ability to sense NO₂ in the presence of other gases (e.g., hydrogen, carbon monoxide, acetylene, ammonia). Combination of metal oxides have also been applied to achieve excellent selectivity and good response time for gases. The sensing properties of metal oxides are attributed to the ease of adsorption/desorption on their surface and consequent changes in electrical properties/conductivities. Carbon nanotubes with increased surface area show excellent sensitivity and very low detection limits as compared to the traditional metal oxide sensors. n-kind graphene exhibits better sensing performance as compared to the p-type. The sensing properties (for target gas) of metal oxides ZnO, SnO_2, In_2O_3, Nb_2O_5 and WO_3 are owed to the redox reaction between metal oxide surface and target gas. The semiconductor sensors have oxide or non-oxide nature. The oxide sensors have the ability to face the harsh conditions and high temperature. Zeolite, perovskites and carbonaceous things have ability to adsorb and detect NO_2. The NO_2 exhibits good adsorption coefficient on the surface of activated carbon. After adsorption, there is an increase of the acidity of AC and decrease of its textural properties. Important NO_2 sensing techniques include selective catalytic reduction (SCR), selective non-catalytic reduction (SNCR) and NO_2 storage reduction (NSR).

References

[1]   C.K. Wilkins, P.A. Clausen, P. Wolkoff, S.T. Larsen, M. Hammer, K. Larsen, V. Hansen, G.D. Nielsen, Formation of strong airway irritants in mixtures of isoprene/ozone and isoprene/ozone/nitrogen dioxide, Environ. Health Perspect. 109 (2001) 937-941. https://doi.org/10.1289/ehp.011099371

[2]   R. Binions, A. Naik, Metal oxide semiconductor gas sensors in environmental monitoring, in Semiconductor gas sensors. 2013, Elsevier. pp. 433-466. https://doi.org/10.1533/9780857098665.4.433

[3]   C. Zhang, Y. Luo, J. Xu, M. Debliquy, Room temperature conductive type metal oxide semiconductor gas sensors for NO_2 detection, Sens. Actuators, A. 289 (2019) 118-133. https://doi.org/10.1016/j.sna.2019.02.027

[4]   S.C. Anenberg, J. Miller, R. Minjares, L. Du, D.K. Henze, F. Lacey, C.S. Malley, L. Emberson, V. Franco, Z. Klimont, Impacts and mitigation of excess diesel-related NOx emissions in 11 major vehicle markets, Nature. 545 (2017) 467-471. https://doi.org/10.1038/nature22086

[5]   G.J. Velders, G.P. Geilenkirchen, R. de Lange, Higher than expected NOx emission from trucks may affect attainability of NO_2 limit values in the netherlands, Atmos. Environ. 45 (2011) 3025-3033. https://doi.org/10.1016/j.atmosenv.2011.03.023

[6]   M.A.H. Khan, M.V. Rao, Q. Li, Recent advances in electrochemical sensors for detecting toxic gases: NO_2, SO_2 and H_2S, Sensors. 19 (2019) 905. https://doi.org/10.3390/s19040905

[7]   A. Afzal, N. Cioffi, L. Sabbatini, L. Torsi, NOx sensors based on semiconducting metal oxide nanostructures: Progress and perspectives, Sens. Actuators, B. 171 (2012) 25-42. https://doi.org/10.1016/j.snb.2012.05.026

[8]   M.A. Bauer, M.J. Utell, P.E. Morrow, D.M. Speers, F.R. Gibb, Inhalation of 0.30 ppm nitrogen dioxide potentiates exercise-induced bronchospasm in asthmatics, Am.Rev. Respir. Dis. 134 (1986) 1203-1208. https://doi.org/10.1164/arrd.1986.134.5.1203

[9]   R. Ehrlich, Effect of nitrogen dioxide on resistance to respiratory infection, Bacter.Rev. 30 (1966) 604-614

[10]   P. Barth, B. Muller, U. Wagner, A. Bittinger, Quantitative analysis of parenchymal and vascular alterations in NO_2-induced lung injury in rats, Eur. Respir. J. 8 (1995) 1115-1121. https://doi.org/10.1183/09031936.95.08071115

[11]   M. Wegmann, A. Fehrenbach, S. Heimann, H. Fehrenbach, H. Renz, H. Garn, U. Herz, NO_2-induced airway inflammation is associated with progressive airflow limitation and development of emphysema-like lesions in C57bl/6 mice, Exp. Toxicol. Pathol. 56 (2005) 341-350. https://doi.org/10.1016/j.etp.2004.12.004

[12]   A. Ponka, M. Virtanen, Chronic bronchitis, emphysema, and low-level air pollution in helsinki, 1987-1989, Environ. Res. 65 (1994) 207-217. https://doi.org/10.1006/enrs.1994.1032

[13]   F.Y. Niyat, I. Sabzevar, M.H.S. Abadi, The review of semiconductor gas sensor for NOx detecting, Turkish Online J. Des. Art Commun. 6 (2016) 898-937. https://doi.org/10.7456/1060JSE/059

[14]   J.A. Bernstein, N. Alexis, C. Barnes, I.L. Bernstein, A. Nel, D. Peden, D. Diaz-Sanchez, S.M. Tarlo, P.B. Williams, Health effects of air pollution, J. Aller.Clinic. Immun. 114 (2004) 1116-1123. https://doi.org/10.1016/j.jaci.2004.08.030

[15]   L. Calderón-Garcidueñas, B. Azzarelli, H. Acuna, R. Garcia, T.M. Gambling, N. Osnaya, S. Monroy, M. Del Rosario Tizapantzi, J.L. Carson, A. Villarreal-Calderon, Air pollution and brain damage, Toxicol. Pathol. 30 (2002) 373-389. https://doi.org/10.1080%2F01926230252929954

[16]   N.M. Elsayed, Toxicity of nitrogen dioxide: An introduction, Toxicol. 89 (1994) 161-174. https://doi.org/10.1016/0300-483X(94)90096-5

[17]   S. Genc, Z. Zadeoglulari, S.H. Fuss, K. Genc, The adverse effects of air pollution on the nervous system, J. Toxicol. 2012 (2012) 1-23. https://doi.org/10.1155/2012/782462

[18]   J. Brunet, V.P. Garcia, A. Pauly, C. Varenne, B. Lauron, An optimised gas sensor microsystem for accurate and real-time measurement of nitrogen dioxide at ppb level, Sens. Actuators, B. 134 (2008) 632-639. https://doi.org/10.1016/j.snb.2008.06.010

[19]   K. Victorin, Review of the genotoxicity of nitrogen oxides, Mut. Res/Rev Gen. Toxicol. 317 (1994) 43-55. https://doi.org/10.1016/0165-1110(94)90011-6

[20]   W. Yan, Y. Yun, T. Ku, G. Li, N. Sang, NO_2 inhalation promotes alzheimer's disease-like progression: Cyclooxygenase-2-derived prostaglandin E-2 modulation and monoacylglycerol lipase inhibition-targeted medication, Sci. Rep. 6 (2016) 22429. https://doi.org/10.1038/srep22429

[21]   R.J. Wild, W.P. Dubé, K.C. Aikin, S.J. Eilerman, J.A. Neuman, J. Peischl, T.B. Ryerson, S.S. Brown, On-road measurements of vehicle NO_2/NOx emission ratios in denver, colorado, USA, Atmos. Environ. 148 (2017) 182-189. https://doi.org/10.1016/j.atmosenv.2016.10.039

[22]   H.E. Stokinger, Evaluation of the hazards of ozone and oxides of nitrogen-factors modifying acute toxicity, J. Air Pollut. Cont. Assoc. 8 (1958) 129-137. https://doi.org/10.1080/00966665.1958.10467839

[23]   D. Zhang, Z. Liu, C. Li, T. Tang, X. Liu, S. Han, B. Lei, C. Zhou, Detection of NO_2 down to ppb levels using individual and multiple In_2O_3 nanowire devices, Nano Lett. 4 (2004) 1919-1924. https://doi.org/10.1021/nl0489283

[24]   R. Atkinson, Atmospheric chemistry of VOCs and NOx, Atmos. Environ. 34 (2000) 2063-2101. https://doi.org/10.1016/S1352-2310(99)00460-4

[25]   T.-Y. Wong, Smog induces oxidative stress and microbiota disruption, J. Food Drug Anal. 25 (2017) 235-244. https://doi.org/10.1016/j.jfda.2017.02.003

[26]   W. Iqbal, Y. Bo, Z. Xu, M. Rauf, M. Waqas, Y. Gong, J, Zhang, Y, Mao,Controllable synthesis of graphitic carbon nitride nanomaterials for solar energy conversion and environmental remediation: The road travelled and the way forward, Catal. Sci. Technol. 8 (2018) 4576-4599. https://doi.org/10.1039/C8CY01061G

[27]   W. Yuan, L. Huang, Q. Zhou, G. Shi, Ultrasensitive and selective nitrogen dioxide sensor based on self-assembled graphene/polymer composite nanofibers, ACS Appl. Mater. Interfaces. 6 (2014) 17003-17008. https://doi.org/10.1021/am504616c

[28]   H.S. Koren, Associations between criteria air pollutants and asthma, Environ. Health Perspect. 103 (1995) 235-242. https://doi.org/10.1289/ehp.95103s6235

[29]   J.Z. Ou, W. Ge, B. Carey, T. Daeneke, A. Rotbart, W. Shan, Y. Wang, Z. Fu, A.F. Chrimes, W. Wlodarski, Physiorption-based charge transfer in two-dimensional SnS_2 for selective and reversible NO_2 gas sensing, ACS Nano. 9 (2015) 10313-10323. https://doi.org/10.1021/acsnano.5b04343

[30]   J. Brunet, M. Dubois, A. Pauly, L. Spinelle, A. Ndiaye, K. Guérin, C. Varenne, B. Lauron, An innovative gas sensor system designed from a sensitive organic

semiconductor downstream a nanocarbonaceous chemical filter for the selective detection of NO_2 in an environmental context: Part I: Development of a nanocarbon filter for the removal of ozone, Sens. Actuators, B. 173 (2012) 659-667. https://doi.org/10.1016/j.snb.2012.07.082

[31]   S. Vilcekova, Indoor nitrogen oxides, in Advanced air pollution. 2011, IntechOpen. https://doi.org/10.5772/16819

[32]   G. Busca, L. Lietti, G. Ramis, F. Berti, Chemical and mechanistic aspects of the selective catalytic reduction of NOx by ammonia over oxide catalysts: A review, Appl. Catal., B. 18 (1998) 1-36. https://doi.org/10.1016/S0926-3373(98)00040-X

[33]   A. Elia, C. Di Franco, A. Afzal, N. Cioffi, L. Torsi, Advanced NOx sensors for mechatronic applications, in Advances in mechatronics. 2011, IntechOpen. https://doi.org/10.5772/22641

[34]   H. Nakamura, I. Haga, K. Murakami, Trend of exhaust emission standards for diesel-powered vehicles, RTRI Report (Railway Technical Research Institute). 20 (2006) 53-56.

[35]   İ.A. Reşitoğlu, NOx pollutants from diesel vehicles and trends in the control technologies, in Diesel engines. 2018, IntechOpen.

[36]   J.W. Erisman, J. Galloway, S. Seitzinger, A. Bleeker, K. Butterbach-Bahl, Reactive nitrogen in the environment and its effect on climate change, Curr. Opin. Environ. Sustain. 3 (2011) 281-290. https://doi.org/10.1016/j.cosust.2011.08.012

[37]   L. Cui, F. Han, W. Dai, E.P. Murray, Influence of microstructure on the sensing behavior of NOx exhaust gas sensors, J. Electrochem. Soc. 161 (2014) B34-B38. https://doi.org/10.1149/2.019403jes

[38]   Y. Huang, Y.S. Yam, C.K. Lee, B. Organ, J.L. Zhou, N.C. Surawski, E.F. Chan, G. Hong, Tackling nitric oxide emissions from dominant diesel vehicle models using on-road remote sensing technology, Environ. Pollut. 243 (2018) 1177-1185. https://doi.org/10.1016/j.envpol.2018.09.088

[39]   A. Font, G.W. Fuller, Did policies to abate atmospheric emissions from traffic have a positive effect in london?, Environ. Pollut. 218 (2016) 463-474. https://doi.org/10.1016/j.envpol.2016.07.026

[40]   M. Akiyama, J. Tamaki, N. Miura, N. Yamazoe, Tungsten oxide-based semiconductor sensor highly sensitive to NO and NO_2, Chem. Lett. 20 (1991) 1611-1614. https://doi.org/10.1246/cl.1991.1611

[41]   N. Yamazoe, G. Sakai, K. Shimanoe, Oxide semiconductor gas sensors, Catal. Surv. Asia. 7 (2003) 63-75.https://doi.org/10.1023/A:1023436725457

[42]   L. Teoh, I. Hung, J. Shich, W. Lai, M.-H. Hon, High sensitivity semiconductor NO_2 gas sensor based on mesoporous WO_3 thin film, Electrochem.Solid-State Lett. 6 (2003) G108-G111. https://doi.org/10.1149/1.1585252

[43]   H. Kurosawa, Y. Yan, N. Miura, N. Yamazoe, Stabilized zirconia-based NOx sensor operative at high temperature, Solid State Ionics. 79 (1995) 338-343. https://doi.org/10.1016/0167-2738(95)00084-J

[44]   L. Wang, Y. Wang, L. Dai, Y. Li, J. Zhu, H. Zhou, High temperature amperometric NO_2 sensor based on nano-structured $Gd_{0.2}Sr_{0.8}Fe_{3-\delta}$ prepared by impregnating method, J. Alloys Compd. 583 (2014) 361-365. https://doi.org/10.1016/j.jallcom.2013.08.168

[45]   S. Fischer, R. Pohle, B. Farber, R. Proch, J. Kaniuk, M. Fleischer, R. Moos, Method for detection of NOx in exhaust gases by pulsed discharge measurements using standard zirconia-based lambda sensors, Sens. Actuators, B. 147 (2010) 780-785. https://doi.org/10.1016/j.snb.2010.03.092

[46]   M. Jeguirim, M. Belhachemi, L. Limousy, S. Bennici, Adsorption/reduction of nitrogen dioxide on activated carbons: Textural properties versus surface chemistry–a review, Chem. Eng. J. 347 (2018) 493-504. https://doi.org/10.1016/j.cej.2018.04.063

[47]   M.J. Kim, K.H. Kim, X. Yang, Y. Yu, Y.S. Lee, Improvement in no gas-sensing properties using heterojunctions between polyaniline and nitrogen on activated carbon fibers, J. Ind. Eng. Chem. 76 (2019) 181-187. https://doi.org/10.1016/j.jiec.2019.03.037

[48]   J. Lee, N. Choi, H. Lee, J. Kim, S. Lim, J. Kwon, S. Lee, S. Moon, J. Jong, D. Yoo, Low power consumption solid electrochemical-type micro CO_2 gas sensor, Sens. Actuators, B. 248 (2017) 957-960. https://doi.org/10.1016/j.snb.2017.02.040

[49]   A. Bandivadekar, K. Bodek, L. Cheah, C. Evans, T. Groode, J. Heywood, E. Kasseris, M. Kromer, M. Weiss, Reducing transportation's petroleum consumption and GHG emissions, (2008).https://doi.org/10.1.1.645.8023

[50]   F. Menil, V. Coillard, C. Lucat, Critical review of nitrogen monoxide sensors for exhaust gases of lean burn engines, Sens. Actuators, B. 67 (2000) 1-23. https://doi.org/10.1016/S0925-4005(00)00401-9

[51]   N. Miura, G. Lu, N. Yamazoe, High-temperature potentiometric/amperometric nox sensors combining stabilized zirconia with mixed-metal oxide electrode, Sens. Actuators, B. 52 (1998) 169-178. https://doi.org/10.1016/S0925-4005(98)00270-6

[52]   J.W. Fergus, Materials for high temperature electrochemical NOx gas sensors, Sens. Actuators, B. 121 (2007) 652-663. https://doi.org/ 10.1016/j.snb.2006.04.077

[53]   J.C. Yang, P.K. Dutta, High temperature amperometric total NOx sensors with platinum-loaded zeolite y electrodes, Sens. Actuators, B. 123 (2007) 929-936. https://doi.org/10.1016/j.snb.2006.10.052

[54]   J. Yoo, F.M. Van Assche, E.D. Wachsman, Temperature-programmed reaction and desorption of the sensor elements of a $WO_3/YSZ/Pt$ potentiometric sensor, J. Electrochem. Soc. 153 (2006) H115-H121. https://doi.org/10.1149/1.2188248

[55]   S.A. Kharitonov, P.J. Barnes, Exhaled markers of pulmonary disease, Am. J. Respir. Crit. Care Med. 163 (2001) 1693-1722. https://doi.org/10.1164/ajrccm.163.7.2009041

[56]   T. Malinski, S. Mesaros, P. Tomboulian, Nitric oxide measurement using electrochemical methods, in Methods in enzymology. 1996, Elsevier. pp. 58-69. https://doi.org/10.1016/S0076-6879(96)68009-4

[57]   A. Fontijn, A.J. Sabadell, R.J. Ronco, Homogeneous chemiluminescent measurement of nitric oxide with ozone. Implications for continuous selective monitoring of gaseous air pollutants, Anal. Chem. 42 (1970) 575-579. https://doi.org/10.1021/ac60288a034

[58]   P.J. Kipping, P. Jeffery, Detection of nitric oxide by gas-chromatography, Nature. 200 (1963) 1314.

[59]   R.M. Palmer, D.D. Rees, D.S. Ashton, S. Moncada, L-arginine is the physiological precursor for the formation of nitric oxide in endothelium-dependent relaxation, Biochem. Biophys. Res. Commun. 153 (1988) 1251-1256. https://doi.org/10.1016/S0006-291X(88)81362-7

[60]   A.A. Kosterev, A.L. Malinovsky, F.K. Tittel, C. Gmachl, F. Capasso, D.L. Sivco, J.N. Baillargeon, A.L. Hutchinson, A.Y. Cho, Cavity ringdown spectroscopic detection of nitric oxide with a continuous-wave quantum-cascade laser, Appl. Opt. 40 (2001) 5522-5529. https://doi.org/10.1364/AO.40.005522

[61]   T. Johnson, Vehicular emissions in review, SAE Int. J. Engines. 6 (2013) 699-715.

[62]   A. Gurlo, N. Barsan, M. Ivanovskaya, U. Weimar, W. Göpel, In_2O_3 and MoO_3–In_2O_3 thin film semiconductor sensors: Interaction with NO_2 and O_3, Sens. Actuators, B. 47 (1998) 92-99. https://doi.org/10.1016/S0925-4005(98)00033-1

[63]   W.F. Bosch, S.A. Dynan, M.D. Shull, Semiconductor processing equipment having improved particle performance. 2003, Google Patents

[64]   N. Barsan, M. Schweizer-Berberich, W. Göpel, Fundamental and practical aspects in the design of nanoscaled SnO_2 gas sensors: A status report, J. Anal. Chem. 365 (1999) 287-304. https://doi.org/10.1007/s002160051490

[65]   N. Yamazoe, J. Fuchigami, M. Kishikawa, T. Seiyama, Interactions of tin oxide surface with O_2, H_2O and H_2, Surf. Sci. 86 (1979) 335-344. https://doi.org/10.1016/0039-6028(79)90411-4

[66]   G. Ko, H.Y. Kim, J. Ahn, Y.M. Park, K.Y. Lee, J. Kim, Graphene-based nitrogen dioxide gas sensors, Curr. Appl. Phys. 10 (2010) 1002-1004. https://doi.org/10.1016/j.cap.2009.12.024

[67]   S. Kannan, H. Steinebach, L. Rieth, F. Solzbacher, Selectivity, stability and repeatability of In_2O_3 thin films towards NOx at high temperatures ($\geq 500°$ c), Sens. Actuators, B. 148 (2010) 126-134. https://doi.org/10.1016/j.snb.2010.04.026

[68]   C. Cantalini, L. Valentini, I. Armentano, L. Lozzi, J. Kenny, S. Santucci, Sensitivity to NO_2 and cross-sensitivity analysis to NH_3, ethanol and humidity of carbon nanotubes thin film prepared by PECVD, Sens. Actuators, B. 95 (2003) 195-202. https://doi.org/10.1016/S0925-4005(03)00418-0

[69]   S. Peng, K. Cho, P. Qi, H. Dai, Ab initio study of CNT NO_2 gas sensor, Chem. Phys. Lett. 387 (2004) 271-276. https://doi.org/10.1016/j.cplett.2004.02.026

[70]   I. Sayago, H. Santos, M. Horrillo, M. Aleixandre, M. Fernández, E. Terrado, I. Tacchini, R. Aroz, W. Maser, A. Benito, Carbon nanotube networks as gas sensors for NO_2 detection, Talanta. 77 (2008) 758-764. https://doi.org/10.1016/j.talanta.2008.07.025

[71]   M. Qazi, T. Vogt, G. Koley, Trace gas detection using nanostructured graphite layers, Appl. Phys. Lett. 91 (2007) 233101. https://doi.org/10.1063/1.2820387

[72]   M.W. Nomani, R. Shishir, M. Qazi, D. Diwan, V. Shields, M. Spencer, G.S. Tompa, N.M. Sbrockey, G. Koley, Highly sensitive and selective detection of NO_2 using epitaxial graphene on 6h-SiC, Sens.Actuators, B. 150 (2010) 301-307. https://doi.org/10.1016/j.snb.2010.06.069

[73]   N. Iqbal, A. Afzal, N. Cioffi, L. Sabbatini, L. Torsi, NOx sensing one-and two-dimensional carbon nanostructures and nanohybrids: Progress and perspectives, Sens. Actuators, B. 181 (2013) 9-21. https://doi.org/10.1016/j.snb.2013.01.089

[74]   Y.T. Ong, A.L. Ahmad, S.H.S. Zein, S.H. Tan, A review on carbon nanotubes in an environmental protection and green engineering perspective, Braz. J. Chem. Eng. 27 (2010) 227-242. https://doi.org/10.1590/S0104-66322010000200002

[75]   M. Penza, D. Suriano, G. Cassano, V. Pfister, M. Alvisi, R. Rossi. Portable chemical sensor-system for urban air-pollution monitoring. in Proceedings of the 14th International Meeting on Chemical Sensors, Nuremberg, Germany. 2012. Citeseer. https://doi.org/10.5162/IMCS2012/P2.9.23

[76]　A. Vaseashta, M. Vaclavikova, S. Vaseashta, G. Gallios, P. Roy, O. Pummakarnchana, Nanostructures in environmental pollution detection, monitoring, and remediation, Sci. Technol. Adv. Mater. 8 (2007) 47. https://doi.org/10.1016/j.stam.2006.11.003

[77]　M. Penza, R. Rossi, M. Alvisi, E. Serra, Metal-modified and vertically aligned carbon nanotube sensors array for landfill gas monitoring applications, Nanotechnol. 21 (2010) 105501

[78]　M.N. Hamidon, Z. Yunusa, P. Wang, Sensing materials for surface acoustic wave chemical sensors, in Progresses in chemical sensor. 2016, InTech. pp. 161-179. https://doi.org/10.5772/63287

[79]　A. Kaushik, R. Khan, V. Gupta, B. Malhotra, S. Ahmad, S. Singh, Hybrid cross-linked polyaniline-WO_3 nanocomposite thin film for NOx gas sensing, J. Nanosci. Nanotechnol. 9 (2009) 1792-1796. https://doi.org/10.1166/jnn.2009.417

[80]　J. Song, Y. Lin, K. Kan, J. Wang, S. Liu, L. Li, K. Shi, Enhanced NOx gas sensing performance based on indium-doped $Co(OH)_2$ nanowire–graphene nanohybrids, Nano. 10 (2015) 1550079. https://doi.org/10.1142/S1793292015500794

[81]　G. Eranna, Metal oxide nanostructures as gas sensing devices. 2016: CRC press.

[82]　E. Llobet, Gas sensors using carbon nanomaterials: A review, Sens. Actuators, B. 179 (2013) 32-45. https://doi.org/10.1142/S1793292015500794

[83]　N.A. Travlou, M. Seredych, E. Rodríguez-Castellón, T.J. Bandosz, Activated carbon-based gas sensors: Effects of surface features on the sensing mechanism, J. Mater. Chem. A. 3 (2015) 3821-3831. https://doi.org/10.1039/C4TA06161F

[84]　R.H. Baughman, A.A. Zakhidov, W.A. De Heer, Carbon nanotubes--the route toward applications, Science. 297 (2002) 787-792. https://doi.org/10.1126/science.1060928

[85]　J. Kong, N.R. Franklin, C. Zhou, M.G. Chapline, S. Peng, K. Cho, H. Dai, Nanotube molecular wires as chemical sensors, Science. 287 (2000) 622-625. https://doi.org/10.1126/science.287.5453.622

[86]　D. Kumar, P. Chaturvedi, P. Saho, P. Jha, A. Chouksey, M. Lal, J. Rawat, R. Tandon, P. Chaudhury, Effect of single wall carbon nanotube networks on gas sensor response and detection limit, Sens. Actuators, B. 240 (2017) 1134-1140. https://doi.org/10.1016/j.snb.2016.09.095

[87]　P.G. Su, C.T. Lee, C.Y. Chou, K.H. Cheng, Y.S. Chuang, Fabrication of flexible NO_2 sensors by layer-by-layer self-assembly of multi-walled carbon nanotubes and

their gas sensing properties, Sens. Actuators B. 139 (2009) 488-493.
https://doi.org/10.1016/j.snb.2016.09.095

[88]   C.P. Fonseca, D.A. Almeida, M.R. Baldan, N.G. Ferreira, NO$_2$ gas sensing using a
CF/PANI composite as electrode, ECS Trans. 41 (2012) 21-28.
https://doi.org/10.1149/1.3695098

[89]   Y.J. Yun, W.G. Hong, N.J. Choi, B.H. Kim, Y. Jun, H.-K. Lee, Ultrasensitive and
highly selective graphene-based single yarn for use in wearable gas sensor, Sci. Rep. 5
(2015) 10904. https://doi.org/10.1038/srep10904

[90]   W. Zhao, C. Yang, D. Zou, Z. Sun, G. Ji, Possibility of gas sensor based on C20
molecular devices, Phys. Lett A. 381 (2017) 1825-1830.
https://doi.org/10.1016/j.physleta.2017.03.038

[91]   W.J. Liou, H.M. Lin, Nanohybrid TiO$_2$/carbon black sensor for NO$_2$ gas, China
Part. 5 (2007) 225-229. https://doi.org/10.1016/j.cpart.2007.03.005

[92]   H. Dai, Carbon nanotubes: Opportunities and challenges, Surf. Sci. 500 (2002)
218-241. https://doi.org/ 10.1016/S0039-6028(01)01558-8

[93]   J. Suehiro, H. Imakiire, S.I. Hidaka, W. Ding, G. Zhou, K. Imasaka, M. Hara,
Schottky-type response of carbon nanotube NO$_2$ gas sensor fabricated onto aluminum
electrodes by dielectrophoresis, Sens. Actuators, B. 114 (2006) 943-949.
https://doi.org/ 10.1016/j.snb.2005.08.043

[94]   S. Liu, B. Yu, H. Zhang, T. Fei, T. Zhang, Enhancing NO$_2$ gas sensing
performances at room temperature based on reduced graphene oxide-ZnO
nanoparticles hybrids, Sens. Actuators,B. 202 (2014) 272-278.
https://doi.org/10.1016/j.snb.2014.05.086

[95]   F.H. Saboor, T. Ueda, K. Kamada, T. Hyodo, Y. Mortazavi, A.A. Khodadadi, Y.
Shimizu, Enhanced NO$_2$ gas sensing performance of bare and Pd-loaded SnO$_2$ thick
film sensors under UV-light irradiation at room temperature, Sens. Actuators, B. 223
(2016) 429-439. https://doi.org/10.1016/j.snb.2015.09.075

[96]   J.J. Adjizian, R. Leghrib, A.A. Koos, I. Suarez-Martinez, A. Crossley, P. Wagner,
N. Grobert, E. Llobet, C.P. Ewels, Boron-and nitrogen-doped multi-wall carbon
nanotubes for gas detection, Carbon. 66 (2014) 662-673.
https://doi.org/10.1016/j.carbon.2013.09.064

[97]   J.L. Figueiredo, Functionalization of porous carbons for catalytic applications, J.
Mater. Chem. A. 1 (2013) 9351-9364. https://doi.org/10.1016/j.carbon.2013.09.064

[98] J.g. Liu, M. Ueda, High refractive index polymers: Fundamental research and practical applications, J. Mater. Chem. 19 (2009) 8907-8919. https://doi.org//10.1039/B909690F

[99] R.-J. Xie, N. Hirosaki, T. Suehiro, F.-F. Xu, M. Mitomo, A simple, efficient synthetic route to $Sr_2Si_5N_8$: Eu^{2+}-based red phosphors for white light-emitting diodes, Chem. Mater. 18 (2006) 5578-5583. https://doi.org/10.1021/cm061010n

[100] K.A. Nielsen, W.S. Cho, G.H. Sarova, B.M. Petersen, A.D. Bond, J. Becher, F. Jensen, D.M. Guldi, J.L. Sessler, J.O. Jeppesen, Supramolecular receptor design: Anion-triggered binding of C60, Angew. Chem. Int. Ed. 45 (2006) 6848-6853. https://doi.org/10.1002/anie.200602724

[101] R. Saito, M. Fujita, G. Dresselhaus, U.M. Dresselhaus, Electronic structure of chiral graphene tubules, Appl. Phys. Lett. 60 (1992) 2204-2206. https://doi.org/10.1063/1.107080@apl.2019.APLCLASS2019.issue-1

[102] B. Seger, P.V. Kamat, Electrocatalytically active graphene-platinum nanocomposites. Role of 2-D carbon support in PEM fuel cells, J. Phys. Chem. C, 113 (2009) 7990-7995. https://doi.org/10.1021/jp900360k

[103] E.J. Biddinger, U.S.J Ozkan, Role of graphitic edge plane exposure in carbon nanostructures for oxygen reduction reaction, J. Phys. Chem. C, 114 (2010) 15306-15314. https://doi.org/10.1021/jp104074t

[104] S.W. Lee, W. Lee, Y. Hong, G. Lee, D.S. Yoon, Recent advances in carbon material-based NO_2 gas sensors, Sens. Actuators, B. 255 (2018) 1788-1804. https://doi.org/10.1016/j.snb.2017.08.203

[105] T. Becker, S. Mühlberger, C.B.V. Braunmühl, G. Müller, T. Ziemann, K. Hechtenberg, Air pollution monitoring using tin-oxide-based microreactor systems, Sens. Actuators, B. 69 (2000) 108-119. https://doi.org/10.1016/S0925-4005(00)00516-5

[106] M. Labaki, M. Issa, S. Smeekens, S. Heylen, C. Kirschhock, K. Villani, M. Jeguirim, D. Habermacher, J. Brilhac, J. Martens, Modeling of NOx adsorption–desorption–reduction cycles on a ruthenium loaded Na–Y zeolite, Appl. Catal., B. 97 (2010) 13-20. https://doi.org/10.1016/S0925-4005(00)00516-5

[107] I. Ghouma, M. Jeguirim, U. Sager, L. Limousy, S. Bennici, E. Däuber, C. Asbach, R. Ligotski, F. Schmidt, A.J.E. Ouederni, The potential of activated carbon made of agro-industrial residues in NOx immissions abatement, Energies, 10 (2017) 1508. https://doi.org/10.3390/en10101508

[108] I. Ghouma, M. Jeguirim, L. Limousy, N. Bader, A. Ouederni, S.J.M. Bennici, Factors influencing NO₂ adsorption/reduction on microporous activated carbon: Poros. Surf. Chem. 11 (2018) 622. https://doi.org/10.3390/ma11040622

[109] M. Jeguirim, M. Belhachemi, L. Limousy, S.J.C.E.J. Bennici, Adsorption/reduction of nitrogen dioxide on activated carbons: Textural properties versus surface chemistry–a review, Chem. Eng. J. 347 (2018) 493-504. https://doi.org/10.1016/j.cej.2018.04.063

[110] U. Sager, W. Schmidt, F.J.A. Schmidt, Catalytic reduction of nitrogen oxides via nanoscopic oxide catalysts within activated carbons at room temperature, Adsorption19 (2013) 1027-1033. https://doi.org/10.1007/s10450-013-9521-8

[111] X. Zhu, L. Zhang, M. Zhang, C.J.F. Ma, Effect of N-doping on NO₂ adsorption and reduction over activated carbon: An experimental and computational study, Fuel. 258 (2019) 116109. https://doi.org/10.1016/j.fuel.2019.116109

[112] M. Jeguirim, V. Tschamber, J. Brilhac, P. Ehrburger, Interaction mechanism of NO₂ with carbon black: Effect of surface oxygen complexes, J. Anal. Appl. Pyrol. 72 (2004) 171-181. https://doi.org/10.1016/j.jaap.2004.03.008

[113] E.M. Suuberg, H. Teng, J.M. Calo. Studies on the kinetics and mechanism of the reaction of no with carbon. in Symposium (International) on combustion. 1991. Elsevier. https://doi.org/10.1016/S0082-0784(06)80381-4

[114] P.A. Lowe, M. Perlsweig. Recent experience for scr systems at coal-fired utility boilers. in Proceedings of the American Power Conference;(United States). 1990

[115] H. Teng, E.M. Suuberg, Chemisorption of nitric oxide on char. 1. Reversible nitric oxide sorption, J. Phys. Chem. 97 (1993) 478-483. https://doi.org/10.1021/j100104a033

[116] M. Belhachemi, M. Jeguirim, L. Limousy, F. Addoun, Comparison of NO₂ removal using date pits activated carbon and modified commercialized activated carbon via different preparation methods: Effect of porosity and surface chemistry, Chem. Eng. J. 253 (2014) 121-129. https://doi.org/10.1016/j.cej.2014.05.004

[117] C.L. Mangun, K.R. Benak, J. Economy, K.L. Foster, Surface chemistry, pore sizes and adsorption properties of activated carbon fibers and precursors treated with ammonia, Carbon. 39 (2001) 1809-1820. https://doi.org/10.1016/S0008-6223(00)00319-5

[118] F. Kapteijn, J. Moulijn, S. Matzner, H.P. Boehm, The development of nitrogen functionality in model chars during gasification in CO₂ and O₂, Carbon. 37 (1999) 1143-1150. https://doi.org/10.1016/S0008-6223(98)00312-1

[119] R. Pietrzak, XPS study and physico-chemical properties of nitrogen-enriched microporous activated carbon from high volatile bituminous coal, Fuel. 88 (2009) 1871-1877. https://doi.org/10.1016/j.fuel.2009.04.017

[120] N.M. Nor, L.C. Lau, K.T. Lee, A.R. Mohamed, Synthesis of activated carbon from lignocellulosic biomass and its applications in air pollution control-a review, J. Environ. Chem. Eng. 1 (2013) 658-666. https://doi.org/10.1016/j.jece.2013.09.017

[121] K. Noll, V. Gounaris, W. Hou, Adsorption technology for air and water pollution control, CRC Press. (1992) pp. 21-22.

[122] R. Pietrzak, T.J. Bandosz, Activated carbons modified with sewage sludge derived phase and their application in the process of NO_2 removal, Carbon. 45 (2007) 2537-2546. https://doi.org/10.1016/j.carbon.2007.08.030

[123] U. Sager, E. Däuber, D. Bathen, C. Asbach, F. Schmidt, J.C. Tseng, A. Pommerin, C. Weidenthaler, W. Schmidt, Influence of the degree of infiltration of modified activated carbons with CuO/ZnO on the separation of NO_2 at ambient temperatures, Adsorpt. Sci. Technol. 34 (2016) 307-319. https://doi.org/10.1177%2F0263617416653120

[124] R. Pietrzak, Sawdust pellets from coniferous species as adsorbents for NO_2 removal, Bioresour. Technol. 101 (2010) 907-913. https://doi.org/10.1016/j.biortech.2009.09.017

[125] P. Nowicki, R. Pietrzak, H. Wachowska, Sorption properties of active carbons obtained from walnut shells by chemical and physical activation, Catal. Today. 150 (2010) 107-114. https://doi.org/10.1016/j.cattod.2009.11.009

[126] P. Nowicki, H. Wachowska, R. Pietrzak, Active carbons prepared by chemical activation of plum stones and their application in removal of NO_2, J. Hazard. Mater. 181 (2010) 1088-1094. https://doi.org/ 10.1016/j.jhazmat.2010.05.12

[127] S. Bashkova, T.J. Bandosz, The effects of urea modification and heat treatment on the process of NO_2 removal by wood-based activated carbon, J. Colloid Interface Sci. 333 (2009) 97-103. https://doi.org/10.1016/j.jcis.2009.01.052

[128] P. Nowicki, P. Skibiszewska, R. Pietrzak, NO_2 removal on adsorbents prepared from coffee industry waste materials, Adsorption. 19 (2013) 521-528. https://doi.org/10.1007/s10450-013-9474-y

[129] K. Pathakoti, M. Manubolu, H.M. Hwang, Nanotechnology applications for environmental industry, in Handbook of nanomaterials for industrial applications. 2018, Elsevier. pp. 894-907. https://doi.org/10.1016/B978-0-12-813351-4.00050-X

[130] G. Korotcenkov, Gas response control through structural and chemical modification of metal oxide films: State of the art and approaches, Sens. Actuators, B. 107 (2005) 209-232. https://doi.org/10.1016/j.snb.2004.10.006

[131] W.T. Moon, K.S. Lee, Y.K. Jun, H.S. Kim, S.H. Hong, Orientation dependence of gas sensing properties of TiO_2 films, Sens. Actuators, B. 115 (2006) 123-127. https://doi.org/10.1016/j.snb.2005.08.024

[132] J. Chang, H. Kuo, I. Leu, M. Hon, The effects of thickness and operation temperature on ZnO: Al thin film co gas sensor, Sens. Actuators, B. 84 (2002) 258-264. https://doi.org/10.1016/S0925-4005(02)00034-5

[133] G. Korotcenkov, The role of morphology and crystallographic structure of metal oxides in response of conductometric-type gas sensors, Mater. Sci. Eng., R. 61 (2008) 1-39. https://doi.org/10.1016/j.mser.2008.02.001

[134] A. Afaah, Z. Khusaimi, M. Rusop. A review on zinc oxide nanostructures: Doping and gas sensing. in Advanced Materials Research. 2013. Trans Tech Publ. https://doi.org/10.4028/www.scientific.net/AMR.667.329.

[135] M. Arafat, B. Dinan, S.A. Akbar, A. Haseeb, Gas sensors based on one dimensional nanostructured metal-oxides: A review, Sensors. 12 (2012) 7207-7258. https://doi.org/10.3390/s120607207

[136] C. Wang, X. Chu, M. Wu, Detection of H_2S down to ppb levels at room temperature using sensors based on zno nanorods, Sens. Actuators, B. 113 (2006) 320-323. https://doi.org/10.1016/j.snb.2005.03.011

[137] Y. Min, H.L. Tuller, S. Palzer, J. Wöllenstein, H. Böttner, Gas response of reactively sputtered ZnO films on si-based micro-array, Sens. Actuators, B. 93 (2003) 435-441. https://doi.org/10.1016/S0925-4005(03)00170-9

[138] Z. Yang, L.M. Li, Q. Wan, Q.H. Liu, T.H. Wang, High-performance ethanol sensing based on an aligned assembly of ZnO nanorods, Sens. Actuators, B. 135 (2008) 57-60. https://doi.org/10.1016/j.snb.2008.07.016

[139] M.H. Huang, S. Mao, H. Feick, H. Yan, Y. Wu, H. Kind, E. Weber, R. Russo, P. Yang, Room-temperature ultraviolet nanowire nanolasers, Science. 292 (2001) 1897-1899. https://doi.org/10.1126/science.1060367

[140] Q. Wan, Q. Li, Y. Chen, T.-H. Wang, X. He, J. Li, C. Lin, Fabrication and ethanol sensing characteristics of ZnO nanowire gas sensors, Appl. Phys. Lett. 84 (2004) 3654-3656. https://doi.org/10.1063/1.1738932@apl.2019.APLCLASS2019.issue-1

[141] L.T.N. Le Viet Thong, N.V.H. Loan, Comparative study of gas sensor performance of SnO_2 nanowires and their hierarchical nanostructures, Sens. Actuators, B. 150 (2010) 112-119. https://doi.org/10.1016/j.snb.2010.07.033

[142] E. Oh, H.Y. Choi, S.H. Jung, S. Cho, J.C. Kim, K.H. Lee, S.W. Kang, J. Kim, J.Y. Yun, S.H. Jeong, High-performance NO_2 gas sensor based on ZnO nanorod grown by ultrasonic irradiation, Sens. Actuators, B. 141 (2009) 239-243. https://doi.org/10.1016/j.snb.2009.06.031

[143] O. Lupan, G. Chai, L. Chow, Novel hydrogen gas sensor based on single ZnO nanorod, Microelectron. Eng. 85 (2008) 2220-2225. https://doi.org/10.1016/j.mee.2008.06.021

[144] Q. Qi, T. Zhang, L. Liu, X. Zheng, Synthesis and toluene sensing properties of SnO_2 nanofibers, Sens. Actuators, B. 137 (2009) 471-475. https://doi.org/10.1016/j.snb.2008.11.042

[145] X. Lu, L. Yin, Porous indium oxide nanorods: Synthesis, characterization and gas sensing properties, J. Mater. Sci.Technol. 27 (2011) 680-684. https://doi.org/10.1016/S1005-0302(11)60125-4

[146] A. Hu, C. Cheng, X. Li, J. Jiang, R. Ding, J. Zhu, F. Wu, J. Liu, X. Huang, Two novel hierarchical homogeneous nanoarchitectures of TiO_2 nanorods branched and P_{25}-coated TiO_2 nanotube arrays and their photocurrent performances, Nanoscale Res. Lett. 6 (2011) 91. https://doi.org/10.1186/1556-276X-6-91

[147] D. Wang, X. Chu, M. Gong, Gas-sensing properties of sensors based on single-crystalline SnO_2 nanorods prepared by a simple molten-salt method, Sens. Actuators, B. 117 (2006) 183-187. https://doi.org/10.1016/j.snb.2005.11.022

[148] Y. Cao, P. Hu, W. Pan, Y. Huang, D. Jia, Methanal and xylene sensors based on ZnO nanoparticles and nanorods prepared by room-temperature solid-state chemical reaction, Sens. Actuators, B. 134 (2008) 462-466. https://doi.org//10.1016/j.snb.2008.05.026

[149] M.H. Huang, Y. Wu, H. Feick, N. Tran, E. Weber, P. Yang, Catalytic growth of zinc oxide nanowires by vapor transport, Adv. Mater. 13 (2001) 113-116. https://doi.org/10.1002/1521-4095(200101)13:2%3C113::AID-ADMA113%3E3.0.CO;2-H

[150] S.S. Kim, J.Y. Park, S.W. Choi, H.S. Kim, H.G. Na, J.C. Yang, H.W. Kim, Significant enhancement of the sensing characteristics of In_2O_3 nanowires by functionalization with Pt nanoparticles, Nanotechnol. 21 (2010) 415502. https://doi.org/10.1088/0957-4484/21/41/415502

[151] C.M. Carney, S. Yoo, S.A. Akbar, TiO_2-SnO_2 nanostructures and their H_2 sensing behavior, Sens. Actuators, B. 108 (2005) 29-33. https://doi.org/10.1016/j.snb.2004.11.058

[152] H.T. Wang, B.S. Kang, F. Ren, L.C. Tien, P. Sadik, D. Norton, S. Pearton, J. Lin, Hydrogen-selective sensing at room temperature with ZnO nanorods, Appl. Phys. Lett. 86 (2005) 243503. https://doi.org/10.1063/1.1949707

[153] C. Baratto, G. Sberveglieri, A. Onischuk, B. Caruso, S. Di Stasio, Low temperature selective NO_2 sensors by nanostructured fibres of ZnO, Sens. Actuators, B. 100 (2004) 261-265. https://doi.org/10.1016/j.snb.2003.12.045

[154] A.Z. Sadek, S. Choopun, W. Wlodarski, S.J. Ippolito, K. Kalantar-zadeh, Characterization of NO_2 nanobelt-based gas sensor forH_2, NO_2, and hydrocarbon sensing, IEEE Sensors Journal. 7 (2007) 919-924. https://doi.org/10.1109/JSEN.2007.895963

[155] N. Zhang, K. Yu, Q. Li, Z. Zhu, Q. Wan, Room-temperature high-sensitivity H_2S gas sensor based on dendritic ZnO nanostructures with macroscale in appearance, J. Appl. Phys. 103 (2008) 104305. https://doi.org/10.1063/1.2924430

[156] Q. Wan, C. Lin, X. Yu, T. Wang, Room-temperature hydrogen storage characteristics of ZnO nanowires, Appl. Phys. Lett. 84 (2004) 124-126. https://doi.org/10.1063/1.1637939

[157] L. Francioso, A. Taurino, A. Forleo, P. Siciliano, TiO_2 nanowires array fabrication and gas sensing properties, Sens. Actuators, B. 130 (2008) 70-76. https://doi.org/10.1016/j.snb.2007.07.074

[158] S. Liu, L. Zhou, L. Yao, L. Chai, L. Li, G. Zhang, K. Shi, One-pot reflux method synthesis of cobalt hydroxide nanoflake-reduced graphene oxide hybrid and their NOx gas sensors at room temperature, J. AlloysCompd. 612 (2014) 126-133. https://doi.org/10.1016/j.jallcom.2014.05.129

[159] A.A. Tomchenko, G.P. Harmer, B.T. Marquis, J.W. Allen, Semiconducting metal oxide sensor array for the selective detection of combustion gases, Sens. Actuators, B. 93 (2003) 126-134. https://doi.org/10.1016/S0925-4005(03)00240-5

[160] V. Bochenkov, G. Sergeev, Preparation and chemiresistive properties of nanostructured materials, Adv. Colloid Interface Sci. 116 (2005) 245-254. https://doi.org/10.1016/j.cis.2005.05.004.

[161] G. Martinelli, M.C. Carotta, M. Ferroni, Y. Sadaoka, E. Traversa, Screen-printed perovskite-type thick films as gas sensors for environmental monitoring, Sens. Actuators, B. 55 (1999) 99-110. https://doi.org/10.1016/S0925-4005(99)00054-4

[162] I. Hotovy, V. Rehacek, P. Siciliano, S. Capone, L. Spiess, Sensing characteristics
of NiO thin films as NO_2 gas sensor, Thin Solid Films. 418 (2002) 9-15.
https://doi.org/10.1016/S0040-6090(02)00579-5

[163] W. Noh, Y. Shin, J. Kim, W. Lee, K. Hong, S.A. Akbar, J. Park, Effects of NiO
addition in WO_3-based gas sensors prepared by thick film process, Solid State Ionics.
152 (2002) 827-832. https://doi.org/10.1016/S0167-2738(02)00341-7

[164] M. Law, H. Kind, B. Messer, F. Kim, P. Yang, Photochemical sensing of NO_2
with SnO_2 nanoribbon nanosensors at room temperature, Angew. Chem. Int. Ed. 41
(2002) 2405-2408. https://doi.org/10.1002/1521-
3773(20020703)41:13%3C2405::AID-ANIE2405%3E3.0.CO;2-3

[165] S. Navale, D. Bandgar, S. Nalage, G. Khuspe, M. Chougule, Y. Kolekar, S. Sen,
V. Patil, Synthesis of Fe_2O_3 nanoparticles for nitrogen dioxide gas sensing
applications, Ceram. Int. 39 (2013) 6453-6460.
https://doi.org/10.1016/j.ceramint.2013.01.074

[166] M. Chougule, S. Sen, V. Patil, Fabrication of nanostructured ZnO thin film sensor
for NO_2 monitoring, Ceram. Int. 38 (2012) 2685-2692.
https://doi.org/10.1016/j.ceramint.2011.11.036

[167] S. Pawar, S. Patil, M. Chougule, B. Raut, S. Pawar, R. Mulik, V. Patil,
Nanocrystalline TiO_2 thin films for NH_3 monitoring: Microstructural and physical
characterization, J. Mater Sci. Mater. Electron. 23 (2012) 273-279.
https://doi.org/10.1007/s10854-011-0403-0

[168] S. Wang, L. Wang, T. Yang, X. Liu, J. Zhang, B. Zhu, S. Zhang, W. Huang, S.
Wu, Porous α-Fe_2O_3 hollow microspheres and their application for acetone sensor, J.
Solid State Chem. 183 (2010) 2869-2876. https://doi.org/10.1016/j.jssc.2010.09.033

[169] L. Huo, Q. Li, H. Zhao, L. Yu, S. Gao, J. Zhao, Sol–gel route to pseudocubic
shaped α-Fe_2O_3 alcohol sensor: Preparation and characterization, Sens. Actuators, B.
107 (2005) 915-920. https://doi.org/10.1016/j.snb.2004.12.046

[170] S. Wang, W. Wang, W. Wang, Z. Jiao, J. Liu, Y. Qian, Characterization and gas-
sensing properties of nanocrystalline iron (iii) oxide films prepared by ultrasonic spray
pyrolysis on silicon, Sens. Actuators, B. 69 (2000) 22-27.
https://doi.org/10.1016/S0925-4005(00)00304-X

[171] E.T. Lee, G.E. Jang, C.K. Kim, D.H. Yoon, Fabrication and gas sensing properties
of α-Fe_2O_3 thin film prepared by plasma enhanced chemical vapor deposition
(PECVD), Sens. Actuators, B. 77 (2001) 221-227. https://doi.org/10.1016/S0925-
4005(01)00716-X

[172] Q. Hao, L. Li, X. Yin, S. Liu, Q. Li, T. Wang, Anomalous conductivity-type transition sensing behaviors of n-type porous α-Fe$_2$O$_3$ nanostructures toward H$_2$S, Mater. Sci.Eng. B. 176 (2011) 600-605. https://doi.org/10.1016/j.mseb.2011.02.002

[173] G. Eranna, B. Joshi, D. Runthala, R. Gupta, Oxide materials for development of integrated gas sensors - A comprehensive review, Crit. Rev. Solid State Mater. Sci. 29 (2004) 111-188. https://doi.org/10.1080/10408430490888977

[174] S. Bai, D. Li, D. Han, R. Luo, A. Chen, C.L. Chung, Preparation, characterization of WO$_3$– SnO$_2$ nanocomposites and their sensing properties for NO$_2$, Sens. Actuators, B. 150 (2010) 749-755. https://doi.org/10.1016/j.snb.2010.08.007

[175] Y.X. Nan, F. Chen, L.G. Yang, H.Z. Chen, Electrochemical synthesis and charge transport properties of CdS nanocrystalline thin films with a conifer-like structure, J. Phys. Chem. C. 114 (2010) 11911-11917. https://doi.org/10.1021/jp103085n

[176] E. Comini, M. Ferroni, V. Guidi, G. Faglia, G. Martinelli, G. Sberveglieri, Nanostructured mixed oxides compounds for gas sensing applications, Sens. Actuators, B. 84 (2002) 26-32. https://doi.org/10.1016/S0925-4005(02)00006-0

[177] R. Ferro, J. Rodriguez, I. Jimenez, A. Cirera, J. Cerda, J. Morante, Gas-sensing properties of sprayed films of CdO$_x$/ZnO$_{1-x}$/mixed oxide, IEEE Sensors Journal. 5 (2005) 48-52. https://doi.org/10.1109/JSEN.2004.838664

[178] K. Galatsis, Y. Li, W. Wlodarski, E. Comini, G. Sberveglieri, C. Cantalini, S. Santucci, M. Passacantando, Comparison of single and binary oxide MoO$_3$, TiO$_2$ and WO$_3$ sol–gel gas sensors, Sens. Actuators, B. 83 (2002) 276-280. https://doi.org/10.1016/S0925-4005(01)01072-3

[179] I.S. Hwang, S.J. Kim, J.K. Choi, J. Choi, H. Ji, G.T. Kim, G. Cao, J.H. Lee, Synthesis and gas sensing characteristics of highly crystalline ZnO– SnO$_2$ core–shell nanowires, Sens. Actuators, B. 148 (2010) 595-600. https://doi.org/10.1016/j.snb.2010.05.052

[180] Z. Jiao, M. Wu, Z. Qin, M. Lu, J. Gu, The NO$_2$ sensing ito thin films prepared by ultrasonic spray pyrolysis, Sensors. 3 (2003) 285-289. https://doi.org/10.3390/s30800285

[181] D.S. Lee, J.W. Lim, S.M. Lee, J.S. Huh, D.D. Lee, Fabrication and characterization of micro-gas sensor for nitrogen oxides gas detection, Sens. Actuators, B. 64 (2000) 31-36. https://doi.org/10.1016/S0925-4005(99)00479-7

[182] I. Murase, A. Moriyama, T. Ito, A. Shimozono, Device for measuring concentration of nitrogen oxide in combustion gas. 1991, Google Patents.

[183] S.J. Hansen, C. HE Burroughs, Managing indoor air quality fifth edition. 2013: Lulu Press, Inc.

[184] X. Kou, C. Wang, M. Ding, C. Feng, X. Li, J. Ma, H. Zhang, Y. Sun, G. Lu, Synthesis of Co-doped SnO_2 nanofibers and their enhanced gas-sensing properties, Sens. Actuators, B. 236 (2016) 425-432. https://doi.org/10.1016/j.snb.2016.06.006

[185] A. Sharma, M. Tomar, V. Gupta, Low temperature operating SnO_2 thin film sensor loaded with WO_3 micro-discs with enhanced response for NO_2 gas, Sens. Actuators, B. 161 (2012) 1114-1118. https://doi.org/10.1016/j.snb.2011.10.014

[186] S.W. Choi, J.Y. Park, S.S. Kim, Synthesis of SnO_2–ZnO core–shell nanofibers via a novel two-step process and their gas sensing properties, Nanotechnol. 20 (2009) 465603. https://doi.org/10.1088/0957-4484/20/46/465603

[187] C. Liangyuan, B. Shouli, Z. Guojun, L. Dianqing, C. Aifan, C.C. Liu, Synthesis of ZnO– SnO_2 nanocomposites by microemulsion and sensing properties for NO_2, Sens. Actuators, B. 134 (2008) 360-366. https://doi.org/10.1016/j.snb.2008.04.040

[188] J.A. Park, J. Moon, S.J. Lee, S.H. Kim, H.Y. Chu, T. Zyung, SnO_2–ZnO hybrid nanofibers-based highly sensitive nitrogen dioxides sensor, Sens. Actuator,s B. 145 (2010) 592-595. https://doi.org/10.1016/j.snb.2009.11.023

[189. N. Yamazoe, N. Miura, Some basic aspects of semiconductor gas sensors, Chem. Sensor Technol. 4 (1992) 19-42.

[190] N. Dirany, Elaboration de matériaux micro-nanostructurés à morphologies contrôlées, à base de tungstates, pour la photo-dégradation. 2017, Toulon

[191] C.Y. Lin, Y.Y. Fang, C.W. Lin, J.J. Tunney, K.C. Ho, Fabrication of NOx gas sensors using In_2O_3-ZnO composite films, Sens. Actuators, B. 146 (2010) 28-34. https://doi.org/10.1016/j.snb.2010.02.040

[192] T. Lee, T. Yun, B. Park, B. Sharma, H.K. Song, B.S. Kim, Hybrid multilayer thin film supercapacitor of graphene nanosheets with polyaniline: Importance of establishing intimate electronic contact through nanoscale blending, J. Mater. Chem. 22 (2012) 21092-21099. https://doi.org/10.1039/C2JM33111J

[193] A. Katoch, Z.U. Abideen, H.W. Kim, S.S. Kim, Grain-size-tuned highly H_2-selective chemiresistive sensors based on ZnO–SnO_2 composite nanofibers, ACS Appl. Mater. Interfaces. 8 (2016) 2486-2494. https://doi.org/10.1021/acsami.5b08416

[194] T.O. Delmont, E. Prestat, K.P. Keegan, M. Faubladier, P. Robe, I.M. Clark, E. Pelletier, P.R. Hirsch, F. Meyer, J.A. Gilbert, Structure, fluctuation and magnitude of a natural grassland soil metagenome, ISME J. 6 (2012) 1677. https://doi.org/10.1038/ismej.2011.197

[195] C. Balázsi, K. Sedlácková, E. Llobet, R. Ionescu, Novel hexagonal WO_3 nanopowder with metal decorated carbon nanotubes as NO_2 gas sensor, Sens. Actuators, B. 133 (2008) 151-155. https://doi.org/10.1016/j.snb.2008.02.006

[196] W. Gomes, S. Lingier, D. Vanmaekelbergh, Anodic stabilization and decomposition mechanisms in semiconductor (photo)-electrochemistry, J. Electroanal. Chem. Interfacial Electrochem. 269 (1989) 237-249. https://doi.org/10.1016/0022-0728(89)85135-6

[197] R. Leghrib, R. Pavelko, A. Felten, A. Vasiliev, C. Cané, I. Gràcia, J.-J. Pireaux, E. Llobet, Gas sensors based on multiwall carbon nanotubes decorated with tin oxide nanoclusters, Sens. Actuators B. 145 (2010) 411-416. https://doi.org/10.1016/j.snb.2009.12.044

[198] F. Kong, Y. Wang, J. Zhang, H. Xia, B. Zhu, Y. Wang, S. Wang, S. Wu, The preparation and gas sensitivity study of polythiophene/SnO_2 composites, Mater. Sci. Eng: B. 150 (2008) 6-11. https://doi.org/10.1016/j.mseb.2008.01.003

Toxic Gas Sensors and Biosensors
Materials Research Foundations **92** (2021) 39-68

Materials Research Forum LLC
https://doi.org/10.21741/9781644901175-2

Chapter 2

Carbon Materials for Gas and Bio-Sensing Applications Beyond Graphene

Ria Majumdar[1], Pinku Chandra Nath[2*]

[1]Department of Civil Engineering, National Institute of Technology, Agartala, Pin-799046

[2]Department of Bio Engineering, National Institute of Technology, Agartala, Pin-799046

* nathpinku005@gmail.com

Abstract

The development of technology in the area of material science and nanotechnology is a worldwide concern to researchers for generating a substance by synthesizing nanoparticles with required properties. Carbonaceous materials have gained numerous interests because of their direct electron or charge transfer capacity between active site reception and functionalized nanoparticles without involvement of a mediator. However, among all existing materials, carbon nanotubes have been proven to elite beyond graphene. Carbon nanotubes (CNTs) possess extraordinary electrochemical biosensing and gas sensing due to their specific properties. This encourages researchers to gain new ideas about construction and development of immunosensors, genosensors, enzymatic biosensors and specific gas sensors based on above nanoparticles. Qualification of working electrode via incorporation of two or more of these nanoparticles gives enhanced stability, better sensitivity and functionality to the sensor. This chapter reviews basic information about sensors, their types, functionalization, fabrication mechanisms and applications for future prospective.

Keywords

Graphene, Biosensors, Gas Sensors, Carbon Nanotubes, Nanomaterials, Metal Oxides

Contents

1.   Introduction

There is a great interest of uncomplicated and authentic gas and biosensors to detect trace elements in a broad range of application. Graphene and carbon nanotubes (CNTs) are two allotropes which have achieved significant research interest to large scale basis applications for the reason of having unique properties. These unique properties include thermal, electronic, mechanical, optical, and transport properties [1]. Graphene and carbon nanotubes (CNTs) have potential applications in industrial, environmental monitoring, transportation, energy, lab-on-a-chip, medical diagnosis, etc. Various experimental as well as theoretical studies proved that sensitivity of the transport and electronic phenomenon of carbonaceous materials are extraordinarily high when it changes in a local chemical environment [2]. Such interference opened up a new and eco-friendly way for the advancement of carbon nanomaterials based gas as well as bio sensors.

Since the last few decades, evolution of biological and chemical sensors made by carbon nanomaterials is an area of recent interest [3]. There are several carbon nanomaterials which have been used for sensing material includes carbon nanotubes (CNTs), nanofibers, nanoparticles, and nanowires. Among all these carbon nanomaterials, carbon nanotubes (CNTs) became most popular sensing materials to detect various gases, trace elements and biocomponents. The sensing materials for detection of gas and bio molecules are potentially used in the field of energy, environment, chemical warfare agents and homeland security [4]. However, the main mechanisms of the sensing material depend on adsorption or desorption of particles and it becomes more effective with the weak binding capacity of substrate (gas or bio) molecules to the surface of carbon nanostructures. The hollow structure of nanomaterials and its extremely high aspect ratio (surface area: volume) are ideal for adsorption of gas molecules as well as gas storage [5]. Therefore, interaction of a non-covalent bonding to covalent bonding is more preferable where bond formation and bond breaking happens simultaneously. It is very important to understand the interaction of non-covalent bonding to the surface of carbon material in order to gain some fundamental and molecular level idea about their different applications [6]. In addition, to understand the behavior and characteristics of such nano-scaled materials, several simulation works and theoretical approaches have been analyzed [7]. There are various factors namely size, curvature of the π-system, etc., which influence non-covalent interaction on the surface of material [8-12]. The adsorption of different substances, for example, gas molecules, metal ions, drugs, biomolecules (i.e. protein and nucleic acids), and organic molecules on the surface of carbon materials have extensively been studied by researchers because of their potential industrial implementations [13].

Recently, the nanotechnology is developing very fast and created a great potential to form ultra sensitive, cost effective, portable sensors with minimum power consumption. It has been shown from the recent studies that reduced graphene oxides (RGO) exhibits an exceptional gas sensing phenomenon [14]. There is a massive demand in several chemical and pharmaceutical industries for highly sensitive sensors to detect leakages of explosive and poisonous gases. In addition, advanced carbon-based nanotechnology has boosted the scope of cost-effective and ultra sensitive sensors for modern daily uses [15].

This current study focuses on a brief study and comparative analysis of non-covalent interaction of different gases, for example, CO_2, H_2O, CH_4, NH_3, H_2 and bio molecules with graphene and carbon nanotubes on the surface of cellulose nanomaterials. The responsiveness of these two allotropes varying with their curvature towards various species is also mentioned [16]. In this chapter, several aspects of graphene and CNTs based gas and biosensors are discussed including the basics of sensors, types, interactions, sensing mechanisms, fabrication process, sensing performance and possible future prospective.

2. Sensors

Definition & backgrounds: Sensors are the devices which can be used for the detection of combustible, noxious gases, flammable, and/or oxygen depletion. Such devices have large potential use in industries, space exploitation, environmental monitoring, biomedicines, pharmaceutics, and also for fire fighting [17].

Requirements: In general, there are several major requirements of a good and efficient sensor to detect: (i) high sensitivity, (ii) high selectivity, (iii) temperature independence, (iv) the first response time, (v) recovery time, (vi) low operating temperature, (vii) less analyst consumption, and (viii) higher stability in performance. Regularly used sensing substances can be made up of semiconductor metal oxides, vapour sensitive polymers, and porous silicon (porous structure materials) [18]. On the other hand, for better enhancement of performance, sensing stuffs are fabricated with different types of substances, for example, optical fibers, conjugated polymers, cyclodextrin, carbon nanomaterials, noble metals, quantum dots and inorganic semiconductor and their derivatives by hybridization [19, 20].

The first biosensor appeared in 1950s with the development of electrochemical devices to detect the analytes. Electrochemical oxygen biosensor is the first among all biosensors which is described by Leland Clerk. It is the most famous biosensor with a platinum cathode in it at which oxygen is reduced as well as a silver/silver chloride reference electrode [22]. This is also known as Clark oxygen electrode. Furthermore, the oxygen electrode is combined with glucose oxidase by Clark and Lyons for measuring the concentration of glucose in solution which is incorporated in a dialysis membrane [23].

In 1967, the first "enzyme electrode" has been introduced by Updike and Hicks. This type of biosensor quantifies the amount of glucose in a solution as well as in tissues *in vitro* which is immobilized with glucose oxidase in a polymerized gelatinous membrane. A polarographic oxygen electrode is coated by this membrane [24] in order to serve as an enzyme electrode which catalyses an electrochemical reaction upon recognition of glucose [23]. The first potentiometric enzyme electrode, also known as urea biosensor, was introduced by Guibault and Montalvo in 1969 according to the immobilization of urease onto an ammonium-selective liquid membrane electrode [25].

A wide range of biosensors are developed for *in vivo* and *in vitro* applications on the basis of their nature. This nature varies from polypeptide, aptamer, to antibody, nucleic-acids based or enzymatic. In this regard, both demand and available technologies are jointly taken into consideration in the mechanism of transduction of biosensors.

Thermal biosensors are developed in 1974 to measure the variation of temperature which is related to the amount of heat released during an enzyme-catalysed reaction. In the following year, i.e. in 1975, the microbial sensors were developed where micro-organisms are integrated with physical transducer (e.g. electrochemical device) for

Materials Research Forum LLC
https://doi.org/10.21741/9781644901175-2

monitoring the specific biomarkers or analytes specially by changing the production of electro-active metabolites or respiration activity.

Immuno biosensors are introduced in 1983 which recognise the target species through antibody fragments or recombinant antibodies. On the other hand, according to the optical diffraction and variation of emitted light signals on target elements, a kind of biosensor has been derived in 1990 which is known as optical biosensor. Recently, in the beginning of the twentieth century, nanobiosensors have been derived based on nanoparticles [26-28]. The development of biosensors over time is shown in Fig.1 [20].

Development:

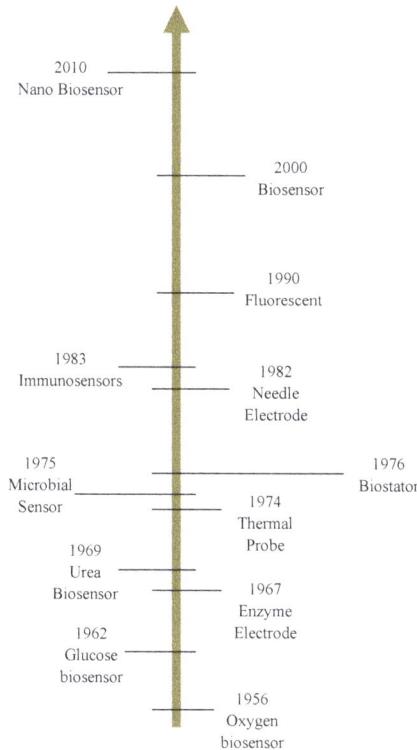

Fig. 1 Development of biosensors over time [21]

3.   Graphene

About: Graphene is a two dimensional nanomaterial which possess unique and attractive sensing properties. It is comprised of atoms of carbon (C) which are closely packed in a hexagonal way to create a monolayer structure. By reason of its unique physicochemical phenomenon, namely aspect ratio (large surface area (SA): volume (V)), potential mechanical durability, good electrical as well as thermal conductivity, graphene has received a considerable amount of interest in the field of sensing. Moreover, single layer graphene sheets possess the largest sensing area per unit volume as all atoms in a single layer acts as a surface. The sensing mechanism of graphene depends on interaction of graphene sheets and adsorbates and this interaction may differ from weak Van der Waals interaction to strong covalent bonding. Thus, its electronic capacity may be affected due to such interactions and can be monitored via convenient electronic techniques [29].

Properties: Graphene is an easily biocompatible, safe and cost effective material. The sensitivity of graphene increases sharply when it is highly loaded with enzymes. Among all the carbon materials known so far, the conductivity of graphene is more as compared to silver. Besides that, it possesses lowest resistivity in the ambient temperature. Generally, it offers a direct transfer of electron as charge between the functionalized graphene and active site of receptor even in the absence of a mediator. Therefore, a noticeable change can be occurred in the conductance of graphene due to a small quantity of excessive electrons. However, the reactivity of graphene electrode can be improved by a conducting polymer such as polyethylenimine (PEI). Moreover, to fabricate four-point devices graphene sheets may be used for effective elimination of the influences of contact resistances [30]. In practical, graphene materials are broadly used to sense the poisonous and explosive gases which are designed precisely to achieve ultra high sensitivity [31].

Reduced graphene oxides (RGO) can be synthesized in large scale by converting graphene materials chemically at relatively low cost. These sheets can be able to assembled or processed into ultra thin sensing layers by applying various wet techniques. To modulate the electronic structure, RGO sheets are functionalized by blending with other sensing materials which can also improve the interactions with gaseous analytes.

3.1   Graphene material based gas sensors

The principal of sensing gas molecules involves adsorption and desorption capacity on the surface of the sensing materials, their hollow structure and large aspect ratio. An experimental study recently stated about high sensitivity graphene-based gas sensors which detect various gases individually [32]. Thus, it is evident that sensitivity of materials can significantly be increased by enhancing the contact interfaces between sensing materials and analytes. To detect leakages of explosive gases for example hydrogen: toxic as well as pathogenic substances in industries, high selectivity and sensitivity based gas sensors are required. Thus, in our recent polluted environment, the capacity for monitoring and controlling ambient environment has gained a numerous demand focusing on the facts of global warming. Detection of different gases with their key parameters using various graphene materials-based gas sensors are illustrated in Table 1 [33].

Table 1   *Detection of different gases with their key parameters using various graphene materials based gas sensors [33]*

Materials	Sensors	Temp. (°C)	Analyte gas	Limit of detection (LOD)	Response time	Ref.
Graphene/Pd	Chemiresistor	-	NO	-	4 min	[117]
Graphene/SnO$_2$	Chemiresistor	260	H$_2$S	1ppm	5s	[118]
Graphene/ZnO	Chemiresistor	340	C$_2$H$_5$OH	100ppm	5s	[119]
Porous graphene	Field Effect Transistors	-	NO$_2$	15 ppb	5–7 min	[120]
Graphene/polypyrene (PPr)	Chemiresistor	-	Toluene	-	10 s	[121]
Chemically modified graphene	Chemiresistor	22-25	NO$_2$	3.6 ppm	5 min	[98]
Mechanically exfoliated graphene	Hall geometry	350	NO$_2$	1 molecule	6 s	[32]
Reduced graphene oxide (RGO)	surface acoustic wave (SAW)	150	H$_2$	200	1 min	[122]
Reduced graphene oxide(RGO)	Chemiresistor	-	Dimethyl Methyl Phosphonate (DMMP)	-	18 min	[123]
Mechanically exfoliated graphene	Chemiresistor	-	CO$_2$	-	8 s	[124]
Epitaxial grown graphene	Chemiresistor	-	NO$_2$	-	5 min	[125]
Chemical Vapor Deposition (CVD) grown graphene	Chemiresistor	-	N$_2$O	103 ppt	2 min	[29]

Chemical Vapor Deposition (CVD) grown graphene	Chemiresistor	-	O_2	38.8 ppt	3 min	[29]
Chemical Vapor Deposition (CVD) growngraphene	Chemiresistor	-	SO_2	67.4 ppt	3 min	[29]
CVD grown graphene	Chemiresistor	-	NO	158 ppt	5 min	[29]
Porous graphene	Field Effect Transistors (FET)	-	NH_3	160 ppb	5-7 min	[120]
Reduced graphene oxide (RGO)	Surface Acoustic Wave (SAW)	-	CO	-	1 min	[126]

3.2   Graphene material based bio sensors

In spite of many detection techniques available with huge money requirement, time consumption, large size, complicated facilities and working principals, and huge number of expert technician requirement, biosensors are still being developed to meet the goals such as high accuracy, quickly functioned, sensitivity in commonly existing reagents and procedures. Graphene was first utilized in 2009 for construction of glucose biosensor. Since then, several graphene-based biosensors are discovered made by incorporation of other substances. Graphene-based biosensors are accessible for dopamine, reduced b-nicotinamide adenine dinucleotide (NADH) molecules, hydrogen peroxides, and enzyme sensing [34-36]. To detect the platelet derived particles in micro scale, a biosensor was designed and fabricated with graphene oxides (GO) [37]. Such biosensors are very simple, easy to operate, high sensitivity in performance i.e. quick, and cost effective. A novel H_2O_2 (hydrogen peroxide) biosensor was developed applying capsules of biomimetic graphene by encapsulation of horseradishperoxidise (HRP) with porous $CaCO_3$ (calcium carbonate) to mimic the real enzymes [38] and can be used as sacrificial templates. A biosensor fabricated with cost-effective graphene for *E.coli* detection gives 60% sensitivity on a flexible acetate sheet against 4.6×10^7cfu per ml concentration of *E.coli* [39]. One more genosensor developed to reveal *E.coli* by the utilization of modified graphene oxide (GO) with iron oxide-chitosen film was newly reported which was deposited on glass substrate coated with indium tin oxide (ITO) [40]. Such biosensors exhibit broad range of detection of 10^{-7}-10^{-15}M and thus 90 percent retention of initial activity may be possible in 6 cycles of use at a time [41]. In addition to this, to modify the glassy carbon electrode immobilized with capture probe, graphene with polythionine and gold nanorod were recently used to detect targeted DNA strand [42]. This medication shows an excellent output in analysis of human serum samples. Therefore, such methods have great potential use in analysis of clinical and diagnostic without the use of enzyme and labeling. Gold particles have capacities to immobilize the immunological entities and graphene itself come up with a large network of active sites. Therefore, graphene fabricated with gold nanoparticles is very useful in assembling immunosensors and senosensors [43, 44]. There is a limitation with enzymatic biosensor when it comes to losses and inactivation of enzyme thereby affecting the performance of sensors. Fabrication of DNA sensors in purpose of specific sequence detection is also a recent interest. Disease causing pathogen detection by such biosensor may build a platform to cure disease and mitigate the problems associated with the pathogens within right time [45]. That's why, construction of novel sensors with accuracy and severe disease sensing capacities is necessary for today's world [46]. Fig. 2 [46] illustrates the generalized approach for construction of graphene-based biosensor.

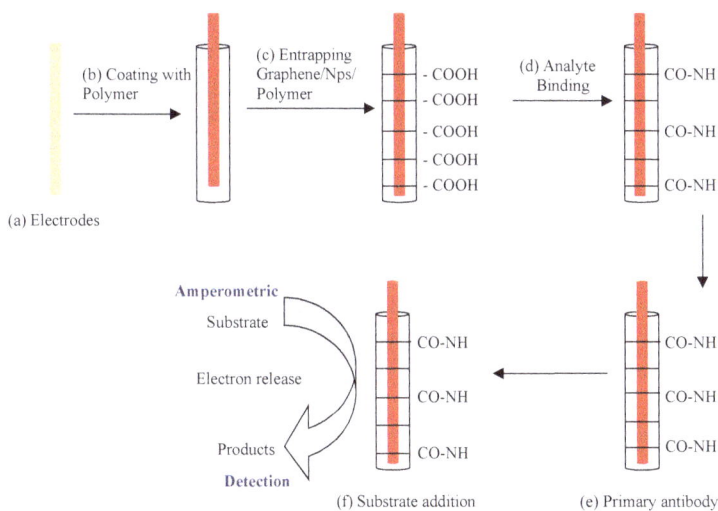

Fig. 2 Graphene-based biosensors (a) Conducting metal electrode (b) Fabricated electrode with polymer coating (c) Improved electrode in combination with graphene (d) Bond formation of analyte with nanomaterials (e) Interaction of analyte and primary antibody (f) Release of electron due to incorporation of substrate leads to amperometric detection.

4. Carbon nanotubes (CNTs)

Historical review: Carbon nanotubes are also familiar as buckytubes [20] and were first discovered by Sumio Iijima, a Japanese electron microscopist, in 1991 [47]. Since his discovery, carbon nanotubes (CNTs) became one of the emerging nano-structured substances which offer new applications to both industrial and academic interest. Millions of papers are being published each year on different aspects of carbon nanomaterials, their properties, fabrication processes and potential applications in different fields.

About CNTs: These belong to the fullerene structures family and are nothing but carbon atoms with sp^2 hybrid structure making them the stiffest and strongest fibers ever. Such nanotubes are hollow from inside and sometimes rolled up with graphene sheets having one or more walls, with diameter in nanoscale and length from nanometres to micrometres [46]. They are having very specific combinations of physical, structural, chemical, mechanical and electronic properties differing from other nanomaterials. CNTs have high mechanical strength due to C-C bond and possess massive thermal stability in both vacuum and air. Furthermore, depending upon the direction of placement of graphite sheets, tube diameter and chirality can be changed thereby making CNTs having good electrical properties (either metallic or semiconducting) [48]. Because of their small size and variable significant conductivity, they have potential applications in the area of molecular electronics to make molecular nanowires. They can be also fabricated with other materials on their surface for better enhancement of the applicability. Both covalent and non-covalent bonding can be used for functionalization of carbon nanotubes with different suitable chemical groups thereby making them biocompatible for unification with bio components. These functional groups namely amine and carboxyl significantly increase the electron transfer rate on the surface of receptor material. Thus, carbon nanotubes can cross the biological barriers effectively for example, the cell membrane, and eventually penetrate into them. If substrates adsorbed on the surface of carbon nanomaterials are capable of modifying the magnetic as well as electronic properties, then presence of various species can be detected by monitoring such specific changes. This specific feature can be a major interest to the researchers for biological and biosensing applications. Any other way, the change in dielectric constants of CNTs selectively detects gas molecules.

Classifications of CNTs: Basically, CNTs are categorized into two types according to their number of walls: (i) Single-walled carbon nanotubes (SWCNTs) and (ii) Multi-walled carbon nanotubes (MWCNTs). However, based on the orientation of tube axis with respect to hexagonal lattice, CNTs can also be split into three more categories: (a) armchair, (b) zigzag, and (c) chiral nanotubes [49-51], which is shown in Fig. 3 [52].

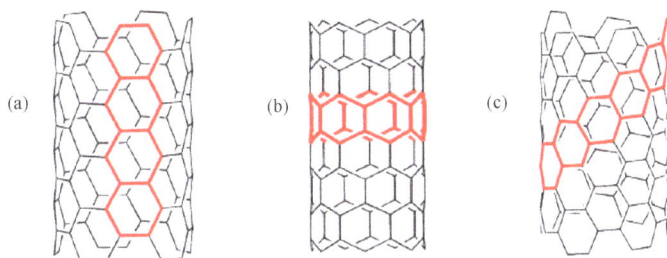

Fig. 3 Classification of CNTs depending on their orientation of axis [52].

SWCNTs are single-atom thick layered rolled up with graphene sheet which possess the simplest morphology. These are cylindrical shaped with diameters in nanometres (or less) and length up to 100 micrometres [53]. Contrarily, MWCNTs are composed of an arrangement of several layers of graphene sheets with diameters of up to 100 nm, sharing the same central axis to make a tube-shaped structure. The distance between two layers in MWCNTs is much less and around 3.5 Å. Moreover, DWCNTs (Double-walled Carbon Nanotubes) are a particular aspect of carbon nanotubes, composed of only two-layered graphene sheets which cumulate the characteristics of both SWCNTs and MWCNTs.

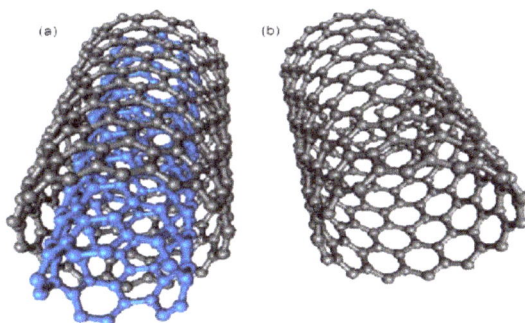

Fig. 4 Classification of carbon nanotubes depending on number of walls (a) MWCNT (Multi-walled carbon nanotube) (b) SWCNT (Single walled carbon nanotube)[54].

4.1 CNTs based gas sensors

Gas sensors are applied in various fields, for example, industrial, medical, and commercial areas. The physical and structural phenomenon of carbon nanotubes (CNTs) make them useful for detecting different gases by adsorbing them on their surfaces. Dai and co-workers [55] were first to describe CNT based gas sensors which detect gases such as NO_2 and NH_3. The literature states that gas molecules are very weak in adsorption on the surface of the carbon nanomaterials due to their low binding energy of about ~5 kcal/mol. The order of this binding energy of gas molecules is as $H_2O > CO_2 > NH_3 > CH_4 > H_2$. This binding property depends on the charge transfer, a property of transferred large electronic charge from charge donor to charge receptor [51]. Therefore, electronic property of the sensor materials is changed. The positive charge value indicates the gas molecules as charge donor to the surface of sensing material. The highest charge transfer occurs in the case of organic vapor or water molecules detection in ambient temperature [56] where H_2 molecules show the lowest electron transfer on the surface. It can be observed from the literatures that charge transfer of different gas molecules to the surface of CNTs depends on chirality of CNTs. But, the graphene material possesses more charge transfer rather than CNTs. Ammonia relates to many environmental problems involving human health, acidification, and change in climate condition by formation of new and harmful particles. Thus, development of ammonia sensors is very much required to monitor the concentration of ambient ammonia [57]. In addition, usage of CNTs based gas sensors are recently reported to detect oxygen [55] and methane [58] for monitoring air quality [57] and controlling industrial processes [59]. For instance, oxygen gas sensors have huge interests in monitoring the environment of combustion engine in order to enhance the performance of engine thereby reducing greenhouse gas emission [60]. Furthermore, CO_2 (carbon dioxide) sensors have broad range of application in medicine and food packages in order to detect spoilage [61, 62]. Among the gas sensors available at market, most of them are operated by estimating the impedance of capacitors incorporated with a gas-sensitive polymer(s) or ceramics namely $CeO/BaCO_3$ (barium carbonate)/CuO (copper oxide) [63], Na_2CO_3 (sodium carbonate) [64], heteropolysiloxane [65], $BaTiO_3$ (barium titanate) [66], Ag_2SO_4 (silver sulfate) [64], SnO_2 (stannic oxide), etc. Though, hard wire connections are required between power supply, sensor head and data processing electronics, such gas sensors possess accuracy with higher degree and reliable performance for better monitoring applications.

4.2 CNTs based biosensors

Proper management and understanding of chemical and physical characteristics of CNTs based biosensors and their surface immobilization and functionalization are very much required for successful realization. The capability of unification of carbon nanotubes (CNTs) with various suitable bio molecules makes them satisfactory substance for biosensing. Carbon nanotubes can be fabricated or immobilized with enzymes to enhance the retentivity of their bio catalytic activity. The combination of enzyme electrodes and specificity of enzymes gives analytical power of electrochemical devices as output makes them extremely suitable for clinical diagnostics and environmental monitoring. This specific feature of carbon nanotubes practically entraps proteins, enzymes, antibodies, DNA, etc. to the surface of the materials by acting as a supporting matrix or scaffold. Clinically important analytes such as cholesterol, lactate, glucose, urate, amino acids, alcohol, hydroxybutyrate, pyruvate, glutamate are catalysed by the enzymes to generate suitable electrochemically products i.e., hydrogen peroxide and NADH [67]. Among all the processes for confining CNT onto electrochemical transducers, CNT/binder composite electrodes [68, 69] and CNT-coated electrodes [70-72] are commonly used. Similar sensitivity and stability improvements have been studied by many researchers for electrochemical biosensor. It is described in Table 2 [73]. Carbon nanotubes (CNTs) can be functionalized with polymers which are water soluble (endohedral functionalization) and also with hydrophilic or ionic groups on surface of material, familiar as exohedral functionalization. These two functionalization processes help CNTs to solubilise in aqueous solution which act as a significant parameter or supporting matrix for scaffolding the entrapment of DNA/proteins/enzymes/antibody. Thus, direct transfer of charge between the active sites and biological element can be improved by functionalizing CNTs. The conductiveness of CNTs specifically MWCNTs are improved by incorporation of hapten molecules, N-ethyl-N-(3-dimethylaminopropyl carbodiimide-N-hydroxysuccinimide (EDC-NHS), redox polymers, and thiol derivatives. It has been also examined that acrylamide helps to improve selectivity and sensitivity of CNTs. The CNTs based sensors possess lower limit detection (0.02mM), long stability, quick response, good reproducibility, and wider linear range. Fig. 5 [46] depicts the simple representation of CNTs based biosensor construction process. The solubilisation of carbon nanomaterials in aqueous media can be determined by both the water-soluble polymers based and functionalized surface with hydrophilic or ionic groups of carbon nanotubes (CNTs).

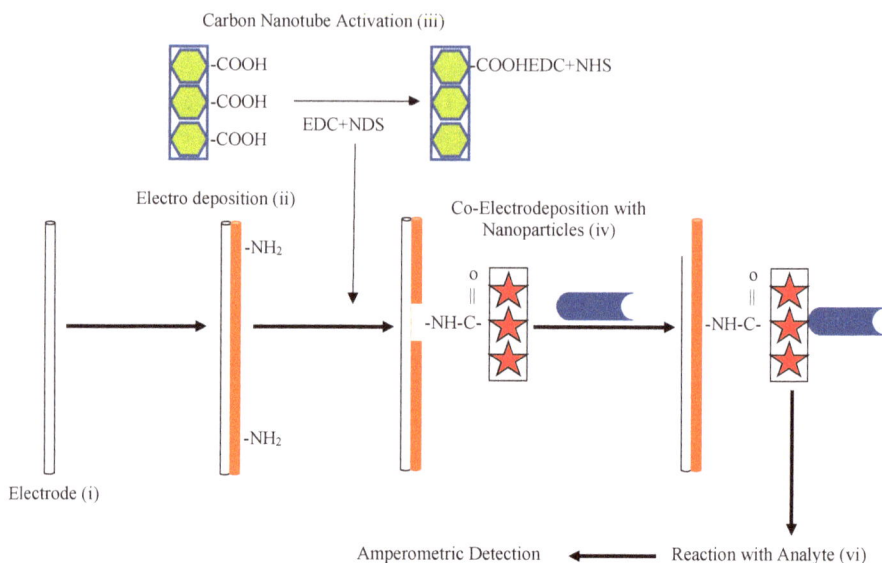

Fig. 5 Carbon nanotubes based biosensors: (i) Metal electrode (ii) Modified electrode (iii) Activation of carboxylated CNTs by EDC-NHS (iv) Co-Electrodeposition of activated CNTs on electrode with nanomaterials (v) Bonding formation between analyte and electrode (vi) Amperometric detection due to reaction with analyte [46].

Table 2   Comparison of analytical performance of CNT/Ni nanocomposite sensor with different non-enzymatic glucose biosensors [73]

Electrode	Applied potential [mV]	Sensitivity [$\mu A.mM^{-1}.cm^{-2}$]	Linear range	Electron transfer rate constant $k_s[s^{-1}]$	Detection limit	Ref.
Ni nanowire arrays	+550 V	1,043	0.5 μM–7 mM	-	0.1 μM	[127]
Pt-Pb/Carbon Nanotube(CNTs)	+300	17.8	Up to 11 mM	-	1.0 μM	[128]

Materials Research Forum LLC
https://doi.org/10.21741/9781644901175-2

Carbon Nanotube(CNT)/glucose oxidase (GOx)	-	2.40	0.04–1.0mM	1.08	-	[129]
Carbon Nanotube(CNT)/Ni	+525	1,433	5 μM–7 mM	-	1 μM	[130]
Porous Au	+350	11.8	2–10 mM	-	5 μM	[131]
Carbon Nanotube(CNT)/polypyrrole/glucose oxidase (GOx)	-	0.095	0.25–4.0mM	-	-	[132]
Carbon Nanotube(CNT)/Colloidal Au/PDDA// glucose oxidase (GOx)	-	2.5×10^3	0.5–5.2mM	1.01	-	[133]
Mesoporous Pt	+400	9.6	0–10 mM	-	N/A	[134]
Nanoporous PtPb	−80	10.8	1–16 mM	-	N/A	[135]
MnO$_2$/ multi-walled carbon nanotubes (MWCNTs)	+300	33.19	10 μM–28 mM	-	10 μM	[136]
Cu/ multi-walled carbon nanotubes (MWCNTs)	+650	251.4	0.7–3.5 mM	-	0.21 μM	[137]
Nafion/ordered mesoporous carbon/glucose oxidase (GOx)	-	0.053	0.5–1.5mM	-	-	[138]

5.   Mechanism of fabrication process of carbon nanotubes (CNTs)

According to integrated dielectric barrier discharge (DBD) principles, a novel ionization gas sensor with short gas spacing, DBD coating and CNTs has been designed and fabricated for realization of breakdown voltage and low power consumption [74]. Fig. 6 [75] shows that inter-digitated electrodes (IDEs) are deposited and patterned on glass substrate. On the other hand, electrophoresis methods are applied for the deposition of carbon nanotubes and then the deposition of a thin film of TiO$_2$ (Titanium dioxide) dielectric barrier was sputtered on the surface of CNT which serves as a DBD layer. It is

Materials Research Forum LLC
https://doi.org/10.21741/9781644901175-2

also evident from the results that various gas sensor fruitfully differentiates through the fingerprinting breakdown voltage [Fig. 7(a)] [75]. Such voltage for air is brought down radically to 5V by a device with 8 μm spacing slit. Thus, the breakdown voltage can be effectively lowered by DBD layer thereby improving the producibility of the device [Fig. 7(b)] [75]. However, low power utilization as well as collapse voltage has given a better opportunity of compact, safe function, battery powered device, and broad area of usages of CNT-based ionization gas sensors.

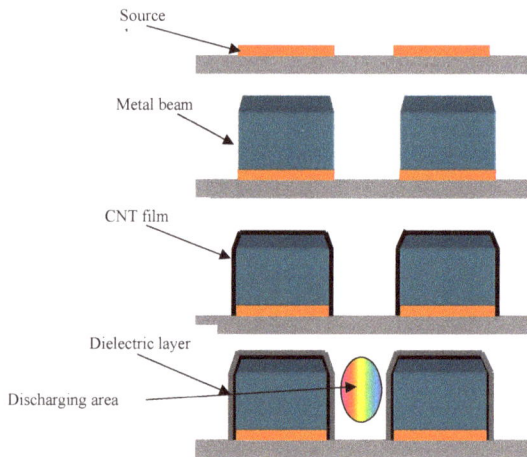

Fig. 6 Fabrication process flow diagram of carbon nanotubes [75].

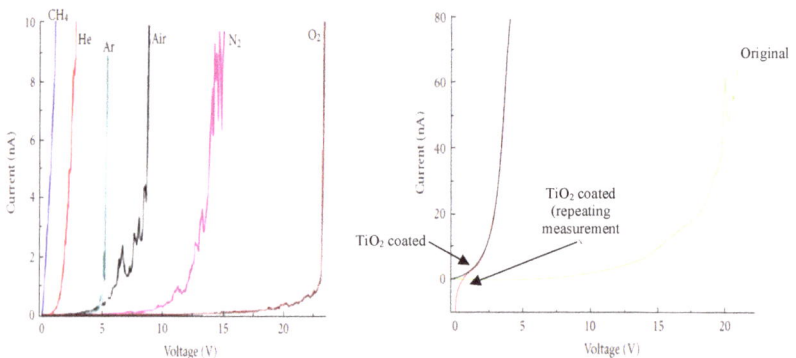

Fig. 7(a) Discharged current-Voltage plots of six various gases (b) gas discharge current –Voltage plots of same device before and after TiO$_2$ coated [75].

6.   Applications of carbon materials and their advantages beyond graphene

CNTs have found use as field emission devices [76], electronic switches [77], actuators [78] and random access memory [79]. Security is the primary constraint of any objects used for biomedical purposes. Carbon nanotubes have many advantages especially in the preservation of electrical and structural characteristics of CNTs. Many researchers are already working using CNTs and their hybrids in applications like clinical chemistry [80, 81], drug delivery [82], fuel cell [83-91], battery [92-94], biodiesel [81], gas sensors [91, 95-104], chemical and biosensors [2, 105-107], greenhouse gas sensor [95], solar cell [88, 92, 108, 107], photonics [109], optical sensor [110], super capacitors [83, 107, 101-105], light-emitting diodes (LEDs) [108], photovoltaic [85, 116], energy, biofuel, optoelectronics, photo catalyst, field effect transistor (FET) and thin-film transistors (Fig. 8) [20].

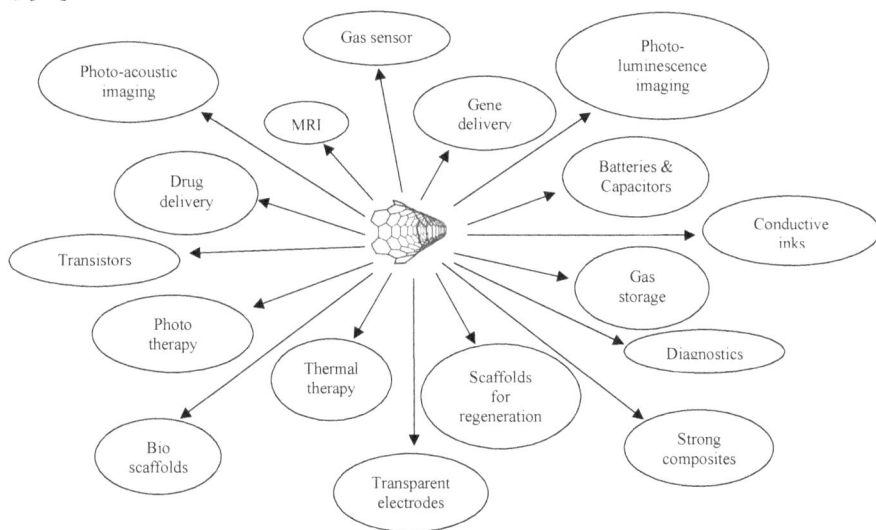

Fig. 8 Application of carbon nanotubes.

Conclusions

Carbon materials have paved the path to make well sensitive sensors with incredible sensing phenomenon. Their sensitivity has given solutions to the unanswered questions. In this chapter, basics of graphene and carbon materials such as CNTs, their fabrication mechanisms and improved sensitivity are discussed. These two materials are used to improve sensitivity to detect analytes with less response time and showed superior reproductivity, longest stability, low detection limit. Such materials also possess exciting opportunities to develop hybrid devices for potential applications in health care. Utilization of above nanomaterials beyond graphene as gas and bio sensors has ended up with a discussion that detection capabilities increase with superior specificity and smaller quantity of analyte.

Biosensors are utilized from a simple blood test to high stage disease treatment like cancer. Carbon nanomaterials based biosensors and gas sensors have potentiality in detecting required components in low range that can sense in pico/femto levels. Thus, such materials have significant advancements in the emerging and developing fields by saving both time and money. In spite of these advantages in the emerging fields, there is still a big challenge to fabricate small size, cheap materials for better enhancement and functionality.

References

[1] W. Yang, K.R. Ratinac, S.P. Ringer, P. Thordarson, J.J. Gooding, F. Braet, Carbon nanomaterials in biosensors: should you use nanotubes or graphene?, Angew. Chem. Int. Ed. 49 (2010) 2114-2138. https://doi.org/ 10.1002/anie.200903463

[2] L. Wang, Q. Zhang, S. Chen, F. Xu, S. Chen, J. Jia, H. Tan, H. Hou, Y. Song, Electrochemical sensing and biosensing platform based on biomass-derived macroporous carbon materials, Anal. Chem. 86 (2014) 1414-1421. https://doi.org/10.1021/ac401563m

[3] S. Mao, G. Lu, J. Chen, Nanocarbon-based gas sensors: progress and challenges, J. Mater. Chem. A 2 (2014) 5573-5579. https://doi.org/ 10.1039/C3TA13823B

[4] D.H. Seo, A.E. Rider, S. Kumar, L.K. Randeniya, K. Ostrikov, Vertical graphene gas- and bio-sensors via catalyst-free, reactive plasma reforming of natural honey, Carbon 60 (2013) 221-228. https://doi.org/ 10.1016/j.carbon.2013.04.015

[5] I. Heller, A.M. Janssens, J. Männik, E.D. Minot, S.G. Lemay, C. Dekker, Identifying the mechanism of biosensing with carbon nanotube transistors, Nano Lett. 8 (2008) 591-595. https://doi.org/ 10.1021/nl072996i

[6] S. Hrapovic, Y. Liu, K.B. Male, J.H. Luong, Electrochemical biosensing platforms using platinum nanoparticles and carbon nanotubes, Anal. Chem. 76 (2004) 1083-1088. https://doi.org/ 10.1021/ac035143t

[7] M. Meyyappan, Carbon nanotubes: science and applications, CRC press2004

[8] D. Vijay, G.N. Sastry, Exploring the size dependence of cyclic and acyclic π-systems on cation–π binding, Phys. Chem. Chem. Phys. 10 (2008) 582-590. https://doi.org/10.1039/B713703F

[9] U.D. Priyakumar, G.N. Sastry, Cation-π interactions of curved polycyclic systems: M+ (M= Li and Na) ion complexation with buckybowls, Tetrahedron Lett. 44 (2003) 6043-6046. https://doi.org/ 10.1016/S0040-4039(03)01512-0

[10] U.D. Priyakumar, M. Punnagai, G.P.K. Mohan, G.N. Sastry, A computational study of cation–π interactions in polycyclic systems: exploring the dependence on the curvature and electronic factors, Tetrahedron 60 (2004) 3037-3043. https://doi.org/ 10.1016/j.tet.2004.01.086

[11] A.S. Mahadevi, G.N. Sastry, Cation− π interaction: Its role and relevance in chemistry, biology, and material science, Chem. Rev. 113 (2013) 2100-2138. https://doi.org/10.1021/cr300222d

[12] M. Chourasia, G.M. Sastry, G.N. Sastry, Aromatic–aromatic interactions database, A²ID: an analysis of aromatic π-networks in proteins, Int. J. Biol. Macromol. 48 (2011) 540-552. https://doi.org/ 10.1016/j.ijbiomac.2011.01.008

[13] J. Ji, J. Wen, Y. Shen, Y. Lv, Y. Chen, S. Liu, H. Ma, Y. Zhang, Simultaneous noncovalent modification and exfoliation of 2D carbon nitride for enhanced electrochemiluminescent biosensing, J. Am. Chem. Soc. 139 (2017) 11698-11701. https://doi.org/ 10.1021/jacs.7b06708

[14] K. Wang, J. Pang, L. Li, S. Zhou, Y. Li, T. Zhang, Synthesis of hydrophobic carbon nanotubes/reduced graphene oxide composite films by flash light irradiation, Front. Chem. Sci. Eng. 12 (2018) 376-382. https://doi.org/ 10.1007/s11705-018-1705-z

[15] Y. Li, J. Yang, Q. Zhao, Y. Li, Dispersing carbon-based nanomaterials in aqueous phase by graphene oxides, Langmuir 29 (2013) 13527-13534. https://doi.org/ 10.1021/la4024025

[16] R. Salahandish, A. Ghaffarinejad, S.M. Naghib, A. Niyazi, K. Majidzadeh-A, M. Janmaleki, A. Sanati-Nezhad, Sandwich-structured nanoparticles-grafted functionalized graphene based 3D nanocomposites for high-performance biosensors to detect ascorbic acid biomolecule, Sci. Rep. 9 (2019) 1-11. https://doi.org/10.1038/s41598-018-37573-9

[17] B. Yuan, C. Xu, L. Liu, Q. Zhang, S. Ji, L. Pi, D. Zhang, Q. Huo, Cu_2O/NiO_x/graphene oxide modified glassy carbon electrode for the enhanced electrochemical oxidation of reduced glutathione and nonenzyme glucose sensor, Electrochim. Acta 104 (2013) 78-83, https://doi.org/ 10.1016/j.electacta.2013.04.073

[18] S. Rumyantsev, G. Liu, M.S. Shur, R.A. Potyrailo, A.A. Balandin, Selective gas sensing with a single pristine graphene transistor, Nano Lett. 12 (2012) 2294-2298. https://doi.org/ 10.1021/nl3001293

[19] K. Xu, C. Fu, Z. Gao, F. Wei, Y. Ying, C. Xu, G. Fu, Nanomaterial-based gas sensors: A review, Instrum. Sci. Technol. 46 (2018) 115-145. https://doi.org/ 10.1080/10739149.2017.1340896

[20] I.V. Pavlidis, M. Patila, U.T. Bornscheuer, D. Gournis, H. Stamatis, Graphene-based nanobiocatalytic systems: recent advances and future prospects, Trends Biotechnol. 32 (2014) 312-320. https://doi.org/ 10.1016/j.tibtech.2014.04.004

[21] C.-M. Tilmaciu, M.C. Morris, Carbon nanotube biosensors, Front. Chem. 3 (2015) 59. https://doi.org/ 10.3389/fchem.2015.00059

[22] L.C. Qlark, Jr., Monitor and control of blood and tissue oxygen tensions, Trans. Am. Soc. Artif. Intern. Organs 2 (1956) 41-48. https://journals.lww.com/asaiojournal/Citation/1956/04000/Monitor_and_Control_of_ Blood_and_Tissue_Oxygen.7.aspx

[23] L.C. Clerk. Jr., and C. Lyons, Electrode systems for continuous monitoring in cardiovascular surgery, Anal. N.Y. Acad. Sci. 102 (1962) 29-45. https://doi.org/ 10.1111/j.1749-6623.1962.tb13623.x

[24] S.J. Updike, G.P. Hicks, The enzyme electrode, Nature 214 (1967) 986-988. https://doi.org/ 10.1038/214986a0

[25] G.G. Guibault, J.G. Montalvo, Urea-specific enzyme electrode, J. Am. Chem. Soc. 91 (1969) 2164-2165. https://doi.org/ 10.1021/ja01036a083

[26] A. Turner, I. Karube, G.S. Wilson, Biosensors: Fundamental and Applications, New York, NY:Oxford University Press, 1987.

[27] M.C. Morris, Fluorescent biosensors of intracellular targets from genetically encoded reporters to modular polypeptide probes, Cell Biochem. Biophys. 56 (2010) 19-37. https://doi.org/ 10.1007/s12013-009-9070-7

[28] A.P.F. Turner, Biosensors: sense and sensitibility, Chem. Soc. Rev. 42 (2013) 3184-3196. https://doi.org/ 10.1039/c3cs35528d

[29] W. Yuan, G. Shi, Graphene-based gas sensors, J. Mater. Chem. A 1 (2013) 10078-10091. https://doi.org/ 10.1039/C3TA11774J

[30] K.C. Kwon, K.S. Choi, B.J. Kim, J.-L. Lee, S.Y. Kim, Work-function decrease of graphene sheet using alkali metal carbonates, J. Phys. Chem. C 116 (2012) 26586-26591. https://doi.org/ 10.1021/jp3069927

[31] Y. Wang, Z. Shi, Y. Huang, Y. Ma, C. Wang, M. Chen, Y. Chen, Supercapacitor devices based on graphene materials, J. Phys. Chem. C 113 (2009) 13103-13107. https://doi.org/ 10.1021/jp902214f

[32] F. Schedin, A.K. Geim, S.V. Morozov, E.W. Hill, P. Blake, M. Katsnelson, K.S. Novoselov, Detection of individual gas molecules adsorbed on graphene, Nat. Mater. 6 (2007) 652-655. https://doi.org/ 10.1038/nmat1967

[33] S.G. Chatterjee, S. Chatterjee, A.K. Ray, A.K. Chakraborty, Graphene-metal oxide nanohybrids for toxic gas sensor: a review, Sens. Actuators B-Chem. 221 (2015) 1170-1181. https://doi.org/ 10.1016/j.snb.2015.07.070

[34] C. Shan, H. Yang, D. Han, Q. Zhang, A. Ivaska, L. Niu, Graphene/AuNPs/chitosan nanocomposites film for glucose biosensing, Biosens. Bioelectron. 25 (2010) 1070-1074. https://doi.org/ 10.1016/j.bios.2009.09.024

[35] Y. Wang, Y. Li, L. Tang, J. Lu, J. Li, Application of graphene-modified electrode for selective detection of dopamine, Electrochem. Commun. 11 (2009) 889-892. https://doi.org/ 10.1016/j.elecom.2009.02.013

[36] S. Alwarappan, A. Erdem, C. Liu, C.-Z. Li, Probing the electrochemical properties of graphene nanosheets for biosensing applications, J. Phys. Chem. C 113 (2009) 8853-8857. https://doi.org/ 10.1021/jp9010313

[37] J. Kailashiya, N. Singh, S.K. Singh, V. Agrawal, D. Dash, Graphene oxide-based biosensor for detection of platelet-derived microparticles: a potential tool for thrombus risk identification, Biosens. Bioelectron. 65 (2015) 274-280. https://doi.org/ 10.1016/j.bios.2014.10.056

[38] Z. Fan, Q. Lin, P. Gong, B. Liu, J. Wang, S. Yang, A new enzymatic immobilization carrier based on graphene capsule for hydrogen peroxide biosensors, Electrochim. Acta 151 (2015) 186-194. https://doi.org/ 10.1016/j.electacta.2014.11.022

[39] P.K. Basu, D. Indukuri, S. Keshavan, V. Navratna, S.R.K. Vanjari, S. Raghavan, N. Bhat, Graphene based E. coli sensor on flexible acetate sheet, Sens. Actuators B-Chem. 190 (2014) 342-347. https://doi.org/ 10.1016/j.snb.2013.08.080

[40] J.M. George, A. Antony, B. Mathew, Metal oxide nanoparticles in electrochemical sensing and biosensing: a review, Microchim. Acta 185 (2018) 358. https://doi.org/ 10.1007/s00604-018-2894-3

[41] I. Tiwari, M. Singh, C.M. Pandey, G. Sumana, Electrochemical genosensor based on graphene oxide modified iron oxide–chitosan hybrid nanocomposite for pathogen detection, Sens. Actuators B-Chem. 206 (2015) 276-283. https://doi.org/ 10.1016/j.snb.2014.09.056

[42] H. Huang, W. Bai, C. Dong, R. Guo, Z. Liu, An ultrasensitive electrochemical DNA biosensor based on graphene/Au nanorod/polythionine for human papillomavirus DNA detection, Biosens. Bioelectron. 68 (2015) 442-446. https://doi.org/ 10.1016/j.bios.2015.01.039

[43] X. Wang, X. Chen, Novel Nanomaterials for Biomedical, Environmental and Energy Applications, Elsevier 2018.

[44] R.D. Pichugov, I.A. Malyshkina, E.E. Makhaeva, Electrochromic behavior and electrical percolation threshold of carbon nanotube/poly (pyridinium triflate) composites, J. Electroanal. Chem. 823 (2018) 601-609. https://doi.org/ 10.1016/j.elechem.2018.07.012

[45] V. Vukojević, S. Djurdjić, M. Ognjanović, M. Fabian, A. Samphao, K. Kalcher, D.M. Stanković, Enzymatic glucose biosensor based on manganese dioxide nanoparticles decorated on graphene nanoribbons, J. Electroanal. Chem. 823 (2018) 610-616. https://doi.org/ 10.1016/j.jelechem.2018.07.013

[46] S. Kumar, W. Ahlawat, R. Kumar, N. Dilbaghi, Graphene, carbon nanotubes, zinc oxide and gold as elite nanomaterials for fabrication of biosensors for healthcare, Biosens. Bioelectron. 70 (2015) 498-503. https://doi.org/ 10.1016/j.bios.2015.03.062

[47] S. Iijima, Helical microtubules of graphitic carbon, Nature 354 (1991) 56-58. https://doi.org/ 10.1038/354056a0

[48] R. Saito, M. Fujita, G. Dresselhaus, M.S. Dresselhaus, Electronic structure of chiral graphene tubules, Appl. Phys. Lett. 60 (1992) 2204-2206. https://doi.org/ 10.1063/1.107080@apl.2019.APLCLASS2019.issue-1

[49] J.Y. Oh, G.H. Jun, S. Jin, H.J. Ryu, S.H. Hong, Enhanced electrical networks of stretchable conductors with small fraction of carbon nanotube/graphene hybrid fillers, ACS Appl. Mater. Interfaces 8 (2016) 3319-3325. https://doi.org/ 10.1021/acsami.5b11205

[50] J.-C. Charlier, Defects in carbon nanotubes, Acc. Chem. Res. 35 (2002) 1063-1069. https://doi.org/ 10.1021/ar010166k

[51] D. Umadevi, G.N. Sastry, Feasibility of carbon nanomaterials as gas sensors: a computational study, 106 (2014) 1224-1234. http://repository.ias.ac.in/108631/1/1224.pdf

[52] T.J, Sisto, L.N. Zakharov, B.M. White, R. Jasti, Towards pi-extended cycloparaphenylenes as seeds for CNT growth: investigating strain relieving ring-openings and rearrangements, Chem. Sci. 7 (2016) 3681-3688. https://doi.org/ 10.1039/C5SC04218F

[53] A.V. Talyzin, I.V. Anoshkin, A.V. Krasheninnikov, R.M. Nieminen, A.G. Nasibulin, H. Jiang, E.I. Kauppinen, Synthesis of graphene nanoribbons encapsulated in single-walled carbon nanotubes, Nano Lett. 11 (2011) 4352-4356. https://doi.org/ 10.1021/nl2024678

[54] M.H. Dehghani, S. Kamalian, M. Shayeghi, M. Yousefi, Z. Heidarinejad, S. Agarwal, V.K. Gupta, High-performance removal of diazinon pesticide from water using multi-walled carbon nanotubes, Microchem. J. 145 (2019) 486-491. https://doi.org/ 10.1016/j.microc.2018.10.053

[55] J. Kong, N.R. Franklin, C. Zhou, M.G. Chapline, S. Peng, K. Cho, H. Dai, Nanotube molecular wires as chemical sensors, Science 287 (2000) 622-625. https://doi.org/ 10.1126/science.287.5453.622

[56] J. Li, Y. Lu, Q. Ye, M. Cinke, J. Han, M. Meyyappan, Carbon nanotube sensors for gas and organic vapor detection, Nano Lett. 3 (2003) 929-933. https://doi.org/ 10.1021/nl034220x

[57] T. Lindgren, D. Norbäck, K. Andersson, B.-G. Dammström, Cabin environment and perception of cabin air quality among commercial aircrew, Aviat. Space Environ. Med. 71 (2000) 774-782. https://europepmc.org/article/med/10954353

[58] M. Bienfait, B. Asmussen, M. Johnson, P. Zeppenfeld, Methane mobility in carbon nanotubes, Surf. Sci. 460 (2000) 243-248. https://doi.org/ 10.1016/S0039-6028(00)00563-X

[59] S.L. Wells, J. DeSimone, CO_2 technology platform: an important tool for environmental problem solving, Angew. Chem. Int. Ed. 40 (2001) 518-527. https://doi.org/ 10.1002/1521-3773(20010202)40:3<518::AID-ANIE518>3.0.CO;2-4

[60] E. Ivers-Tiffée, K.H. Härdtl, W. Menesklou, J. Riegel, Principles of solid state oxygen sensors for lean combustion gas control, Electrochim. Acta 47 (2001) 807-814. https://doi.org/ 10.1016/S0013-4686(01)00761-7

[61] N.F. Sheppard Jr, R.C. Tucker, S. Salehi-Had, Design of a conductimetric pH microsensor based on reversibly swelling hydrogels, Sens. Actuators B-Chem. 10 (1993) 73-77. https://doi.org/ 10.1016/0925-4005(93)80028-A

[62] P. Dalgaard, O. Mejlholm, H.H. Huss, Application of an iterative approach for development of a microbial model predicting the shelf-life of packed fish, Int. J. Food Microbiol. 38 (1997) 169-179. https://doi.org/ 10.1016/S0168-1605(97)00101-3

[63] S. Matsubara, S. Kaneko, S. Morimoto, S. Shimizu, T. Ishihara, Y. Takita, A practical capacitive type CO_2 sensor using $CeO_2/BaCO_3/CuO$ ceramics, Sens. Actuators B-Chem. 65 (2000) 128-132. https://doi.org/ 10.1016/S0925-4005(99)00407-4

[64] J.F. Currie, A. Essalik, J-C Marusic, Micromachined thin film solid state electrochemical CO_2, NO_2 and SO_2 gas sensors, Sens. Actuators B-Chem. 59 (1999) 235-241. https://doi.org/ 10.1016/S0925-4005(99)00227-0

[65] H.-E. Endres, R. Hartinger, M. Schwaiger, G. Gmelch, M. Roth, A capacitive CO_2 sensor system with suppression of the humidity interference, Sens. Actuators B-Chem. 57 (1999) 83-87. https://doi.org/ 10.1016/S0925-4005(99)00060-X

[66] M.-S. Lee, J.-U. Meyer, A new process for fabricating CO_2-sensing layers based on $BaTiO_3$ and additives, Sens. Actuators B-Chem. 68 (2000) 293-299. https://doi.org/ 10.1016/S0925-4005(00)00447-0

[67] J. Wang, Carbon-nanotube based electrochemical biosensors: A review, Electroanalysis (N.Y.N.Y.) 17 (2005) 7-14. https://doi.org/ 10.1002/elan.200403113

[68] J. Wang, M. Musameh, Carbon nanotube/teflon composite electrochemical sensors and biosensors, Anal. Chem. 75 (2003) 2075-2079. https://doi.org/ 10.1021/ac030007+

[69] M. D. Rubianes, G.A. Rivas, Carbon nanotubes paste electrode, Electrochem. Commun. 5 (2003) 689-694. https://doi.org/ 10.1016/S1388-2481(03)00168-1

[70] M. Musameh, J. Wang, A. Merkoci, Y. Lin, Low-potential stable NADH detection at carbon-nanotube-modified glassy carbon electrodes, Electrochem. Commun. 4 (2002) 743-746. https://doi.org/ 10.1016/S1388-2481(02)00451-4

[71] J. Wang, M. Musameh, Y. Lin, Solubilization of carbon nanotubes by nafion toward the preparation of amperometric biosensors, J. Am. Chem. Soc. 125 (2003) 2408-2409. https://doi.org/ 10.1021/ja028951v

[72] J.H.T. Luong, S. Hrapovic, D. Wang, F. Bensebaa, B. Simard, Solubilization of multiwall carbon nanotubes by 3-aminopropyltriethoxysilane towards the fabrication of electrochemical biosensors with promoted electron transfer, Electroanalysis (N.Y.N.Y.) 16 (2004) 132-139. https://doi.org/ 10.1002/elan.200302931

[73] S. Gupta, C.R. Prabha, C.N. Murthy, Functionalized multi-walled carbon nanotubes/polyvinyl alcohol membrane coated glassy carbon electrode for efficient enzyme immobilization and glucose sensing, J. Environ. Chem. Eng. 4 (2016) 3734-3740. https://doi.org/ 10.1016/j.jece.2016.08.021

[74] C.-S. Woo, C.-H. Lim, C.-W. Cho, B. Park, H. Ju, D.-H. Min, C.-J. Lee, S.-B. Lee, Fabrication of flexible and transparent single-wall carbon nanotube gas sensors by vacuum filtration and poly (dimethyl siloxane) mold transfer, Microelectron. Eng. 84 (2007) 1610-1613. https://doi.org/ 10.1016/j.mee.2007.01.162

[75] Y. Wang, J.T.W. Yeow, A review of carbon nanotubes-based gas sensors, J. Sens. 2009 (2009). https://doi.org/ 10.1155/2009/493904

[76] W.A. de Heer, A. Chatelain, D. Ugarte, A carbon nanotube field-emission electron source, Science 270 (1995) 1179-1180. https://doi.org/10.1126/science.270.5239.1179

[77] S.J. Tans, A.R.M. Verschueren, C. Dekker, Room-temperature transistor based on a single carbon nanotube, Nature 393 (1998) 49-52. https://doi.org/ 10.1038/29954

[78] R.H. Baughman, C. Cui, A.A. Zakhidov, Z. Iqbal, J.N. Barisci, G.M. Spinks, G.G. Wallace, A. Mazzoldi, D. De Rossi, A.G. Rinzler, O. Jaschinski, S. Roth, M. Kertesz, Carbon nanotube actuators, Science 284 (1999) 1340-1344. https://doi.org/ 10.1126/science.284.5418.1340

[79] T. Rueckes, K. Kim, E. Joselevich, G.Y. Tseng, C.-L. Cheung, C.M. Lieber, Carbon nanotube-based nonvolatile random access memory for molecular computing, Science 289 (2000) 94-97. https://doi.org/ 10.1126/science.289.5476.94

[80] A.M. Pisoschi, Biosensors as bio-based materials in chemical analysis: a review, J. Biobased Mater. Bioenergy 7 (2013) 19-38. https://doi.org/ 10.1166/jbmb.2013.1274

[81] S.K. Vashist, D. Zheng, K. Al-Rubeaan, J.H.T. Luong, F.-S. Sheu, Advances in carbon nanotube based electrochemical sensors for bioanalytical applications, Biotechnol. Adv. 29 (2011) 169-188. https://doi.org/ 10.1016/j.biotechadv.2010.10.002

[82] J. Kirsch, C. Siltanen, Q. Zhou, A. Revzin, A. Simonian, Biosensor technology: recent advances in threat agent detection and medicine, Chem. Soc. Rev. 42 (2013) 8733-8768. https://doi.org/ 10.1039/C3CS60141B

[83] F. Xiaomiao, L. Ruimei, Y. Xiaoyan, H. Wenhua, Application of novel carbon nanomaterials to electrochemistry [J], Prog. Chem. 11 (2012). http://en.cnki.com.cn/Article_en/CJFDTotal-HXJZ201211010.htm

[84] D.J. Caruana, S. Howorka, Biosensors and biofuel cells with engineered proteins, Mol. Biosyst. 6 (2010) 1548-1556. https://doi.org/ 10.1039/C004951D

[85] S. Zhu, G. Xu, Single-walled carbon nanohorns and their applications, Nanoscale 2 (2010) 2538-2549. https://doi.org/ 10.1039/C0NR00387E

[86] W. Feng, P. Ji, Enzymes immobilized on carbon nanotubes, Biotechnol. Adv. 29 (2011) 889-895. https://doi.org/ 10.1016/j.biotechadv.2011.07.007

[87] T. Nöll, G. Nöll, Strategies for "wiring" redox-active proteins to electrodes and applications in biosensors, biofuel cells, and nanotechnology, Chem. Soc. Rev. 40 (2011) 3564-3576. https://doi.org/ 10.1039/C1CS15030H

[88] Q. Lang, L. Yin, J. Shi, L. Li, L. Xia, A. Liu, Co-immobilization of glucoamylase and glucose oxidase for electrochemical sequential enzyme electrode for starch biosensor and biofuel cell, Biosens. Bioelectron. 51 (2014) 158-163. https://doi.org/ 10.1016/j.bios.2013.07.021

[89] X.-Y. Yang, G. Tian, N. Jiang, B.-L. Su, Immobilization technology: a sustainable solution for biofuel cell design, Energy Environ. Sci. 5 (2012) 5540-5563. https://doi.org/ 10.1039/C1EE02391H

[90] Y. Tan, W. Deng, B. Ge, Q. Xie, J. Huang, S. Yao, Biofuel cell and phenolic biosensor based on acid-resistant laccase–glutaraldehyde functionalized chitosan–multiwalled carbon nanotubes nanocomposite film, Biosens. Bioelectron. 24 (2009) 2225-2231. https://doi.org/ 10.1016/j.bios.2008.11.026

[91] S.H. Shuit, K.F. Yee, K.T. Lee, B. Subhash, S.H. Tan, Evolution towards the utilisation of functionalised carbon nanotubes as a new generation catalyst support in biodiesel production: an overview, RSC Adv. 3 (2013) 9070-9094. https://doi.org/ 10.1039/C3RA22945A

[92] Y. Ye, C. Jo, I. Jeong, J. Lee, Functional mesoporous materials for energy applications: solar cells, fuel cells, and batteries, Nanoscale 5 (2013) 4584-4605. https://doi.org/ 10.1039/C3NR00176H

[93] S. Zhang, Y. Shao, G. Yin, Y. Lin, Recent progress in nanostructured electrocatalysts for PEM fuel cells, J. Mater. Chem. A 1 (2013) 4631-4641. https://doi.org/ 10.1039/C3TA01161E

[94] Q. Li, R. Cao, J. Cho, G. Wu, Nanostructured carbon-based cathode catalysts for nonaqueous lithium–oxygen batteries, Phys. Chem. Chem. Phys. 16 (2014) 13568-13582. https://doi.org/ 10.1039/C4CP00225C

[95] D. Olney, L. Fuller, K.S.V. Santhanam, A greenhouse gas silicon microchip sensor using a conducting composite with single walled carbon nanotubes, Sens. Actuators B-Chem. 191 (2014) 545-552. https://doi.org/ 10.1016/j.snb.2013.10.039

[96] S. Dhall, N. Jaggi, R. Nathawat, Functionalized multiwalled carbon nanotubes based hydrogen gas sensor, Sens. Actuators A Phys. 201 (2013) 321-327. https://doi.org/ 10.1016/j.sna.2013.07.018

[97] S. Badhulika, N.V. Myung, A. Mulchandani, Conducting polymer coated single-walled carbon nanotube gas sensors for the detection of volatile organic compounds, Talanta 123 (2014) 109-114. https://doi.org/ 10.1016/j.talanta.2014.02.005

[98] W. Yuan, L. Huang, Q. Zhou, G. Shi, Ultrasensitive and selective nitrogen dioxide sensor based on self-assembled graphene/polymer composite nanofibers, ACS Appl. Mater. Interfaces 6 (2014) 17003-17008. https://doi.org/ 10.1021/am504616c

[99] M. Mittal, A. Kumar, Carbon nanotube (CNT) gas sensors for emissions from fossil fuel burning, Sens. Actuators B-Chem. 203 (2014) 349-362. https://doi.org/ 10.1016/j.snb.2014.05.080

[100] E. Akbari, Z. Buntat, A. Enzevaee, M. Ebrahimi, A.H. Yazdavar, R. Yusof, Analytical modeling and simulation of I–V characteristics in carbon nanotube based gas sensors using ANN and SVR methods, Chemom. Intell. Lab. Syst. 137 (2014) 173-180. https://doi.org/ 10.1016/j.chemolab.2014.07.001

[101] P. Jha, M. Sharma, A. Chouksey, P. Chaturvedi, D. Kumar, G. Upadhyaya, J. S.B.S. Rawat, P. K. Chaudhury, Functionalization of carbon nanotubes with metal phthalocyanine for selective gas sensing application, Synth. React. Inorg. Met-org. Nano-met. Chem. 44 (2014) 1551-1557. https://doi.org/ 10.1080/15533174.2013.818021

[102] L. Lvova, M. Mastroianni, G. Pomarico, M. Santonico, G. Pennazza, C. Di Natale, R. Paolesse, A. D'Amico, Carbon nanotubes modified with porphyrin units for gaseous phase chemical sensing, Sens. Actuators B-Chem. 170 (2012) 163-171. https://doi.org/ 10.1016/j.snb.2011.05.031

[103] F. Xu, S. Guo, Y.-L. Luo, Novel THTBN/MWNTs-OH polyurethane conducting composite thin films for applications in detection of volatile organic compounds, Mater. Chem. Phys. 145 (2014) 222-231. https://doi.org/ 10.1016/j.matchemphys.2014.02.006

[104] D. Jung, M. Han, G.S. Lee, Gas-sensing properties of multi-walled carbon-nanotube sheet coated with NiO, Carbon 78 (2014) 156-163. https://doi.org/ 10.1016/j.carbon.2014.06.063

[105] M. Boujtita, Chemical and biological sensing with carbon nanotubes (CNTs), Nanosens. Chem. Biol. App. (2014) 3-27. https://doi.org/ 10.1533/9780857096722.1.3

[106] M. Ates, A.S. Sarac, Conducting polymer coated carbon surfaces and biosensor applications, Prog. Org. Coat. 66 (2009) 337-358. https://doi.org/ 10.1016/j.porgcoat.2009.08.014

[107] H.J. Salavagione, A.M. Díez-Pascual, E. Lázaro, S. Vera, M.A. Gómez-Fatou, Chemical sensors based on polymer composites with carbon nanotubes and graphene: the role of the polymer, J. Mater. Chem. A 2 (2014) 14289-14328. https://doi.org/ 10.1039/C4TA02159B

[108] S. Park, M. Vosguerichian, Z. Bao, A review of fabrication and applications of carbon nanotube film-based flexible electronics, Nanoscale 5 (2013) 1727-1752. https://doi.org/ 10.1039/C3NR33560G

[109] M.A. Arugula, A. Simonian, Novel trends in affinity biosensors: current challenges and perspectives, Meas. Sci. Technol. 25 (2014) 032001. https://doi.org/ 10.1088/0957-0233/25/3/032001

[110] S. Kruss, A.J. Hilmer, J. Zhang, N.F. Reuel, B. Mu, M.S. Strano, Carbon nanotubes as optical biomedical sensors, Adv. Drug Deliv. Rev. 65 (2013) 1933-1950. https://doi.org/ 10.1016/j.addr.2013.07.015

[111] V. Mani, S.-M. Chen, B.-S. Lou, Three dimensional graphene oxide-carbon nanotubes and graphene-carbon nanotubes hybrids, Int. J. Electrochem. Sci. 8 (2013) 11641-11660. http://electrochemsci.org/papers/vol8/81011641.pdf

[112] K. Ryu, H. Xue, J. Park, Benign enzymatic synthesis of multiwalled carbon nanotube composites uniformly coated with polypyrrole for supercapacitors, J. Chem. Technol. Biotechnol. 88 (2013) 788-793. https://doi.org/ 10.1002/jctb.3899

[113] X. Zhao, B.M. Sánchez, P.J. Dobson, P.S. Grant, The role of nanomaterials in redox-based supercapacitors for next generation energy storage devices, Nanoscale 3 (2011) 839-855. https://doi.org/ 10.1039/C0NR00594K

[114] J. Yang, L. Lian, P. Xiong, M. Wei, Pseudo-capacitive performance of titanate nanotubes as a supercapacitor electrode, Chem. Commun. 50 (2014) 5973-5975. https://doi.org/ 10.1039/C3CC49494B

[115] M. Sawangphruk, M. Suksomboon, K. Kongsupornsak, J. Khuntilo, P. Srimuk, Y. Sanguansak, P. Klunbud, P. Suktha, P. Chiochan, High-performance supercapacitors based on silver nanoparticle–polyaniline–graphene nanocomposites coated on flexible carbon fiber paper, J. Mater. Chem. A 1 (2013) 9630-9636. https://doi.org/ 10.1039/C3TA12194A

[116] S. Cataldo, P. Salice, E. Menna, B. Pignataro, Carbon nanotubes and organic solar cells, Energy Environ. Sci. 5 (2012) 5919-5940. https://doi.org/ 10.1039/C1EE02276H

[117] W. Yang, L. Gan, H. Li, T. Zhai, Two-dimensional layered nanomaterials for gas-sensing applications, Inorg. Chem. Front. 3 (2016) 433-451. https://doi.org/ 10.1039/C5QI00251F

[118] Z. Zhang, R. Zou, G. Song, L. Yu, Z. Chen, J. Hu, Highly aligned SnO_2 nanorods on graphene sheets for gas sensors, J. Mater. Chem. 21 (2011) 17360-17365. https://doi.org/ 10.1039/C1JM12987B

[119] J. Yi, J.M. Lee, W.I. Park, Vertically aligned ZnO nanorods and graphene hybrid architectures for high-sensitive flexible gas sensors, Sens. Actuators B-Chem. 155 (2011) 264-269. https://doi.org/ 10.1016/j.snb.2010.12.033

[120] T.H. Han, Y.-K. Huang, A.T.L. Tan, V.P. Dravid, J. Huang, Steam etched porous graphene oxide network for chemical sensing, J. Am. Chem. Soc. 133 (2011) 15264-15267. https://doi.org/ 10.1021/ja205693t

[121] L. Zhang, C. Li, A. Liu, G. Shi, Electrosynthesis of graphene oxide/polypyrene composite films and their applications for sensing organic vapors, J. Mater. Chem. 22 (2012) 8438-8443. https://doi.org/ 10.1039/C2JM16552J

[122] R. Arsat, M. Breedon, M. Sha, ei, PG Spizziri, S. Gilje, RB Kaner, K. Kalantar-zadeh and W. Wlodarski, Chem. Phys. Lett. 467 (2009) 344-347

[123] N. Hu, Y. Wang, J. Chai, R. Gao, Z. Yang, E.S.-W. Kong, Y. Zhang, Gas sensor based on p-phenylenediamine reduced graphene oxide, Sens. Actuators B-Chem. 163 (2012) 107-114. https://doi.org/ 10.1016/j.snb.2012.01.016

[124] H.J. Yoon, D.H. Jun, J.H. Yang, Z. Zhou, S.S. Yang, M.M.-C. Cheng, Carbon dioxide gas sensor using a graphene sheet, Sens. Actuators B-Chem. 157 (2011) 310-313. https://doi.org/ 10.1016/j.snb.2011.03.035

[125] M.W.K. Nomani, R. Shishir, M. Qazi, D. Diwan, V. B. Shields, M. G. Spencer, G.S. Tompa, N.M. Sbrockey, G. Koley, Highly sensitive and selective detection of NO_2 using epitaxial graphene on 6H-SiC, Sens. Actuators B-Chem. 150 (2010) 301-307. https://doi.org/ 10.1016/j.snb.2010.06.069

[126] R. Arsat, M. Breedon, M. Shafiei, P.G. Spizziri, S. Gilje, R. B. Kaner, K. Kalantar-zadeh, W. Wlodarski, Graphene-like nano-sheets for surface acoustic wave gas sensor applications, Chem. Phys. Lett. 467 (2009) 344-347. https://doi.org/ 10.1016/j.cplett.2008.11.039

[127] L.-M. Lu, L. Zhang, F.-L. Qu, H.-X. Lu, X.-B. Zhang, Z.-S. Wu, S.-Y. Huan, Q.-A. Wang, G.-L. Shen, R.-Q. Yu, A nano-Ni based ultrasensitive nonenzymatic electrochemical sensor for glucose: enhancing sensitivity through a nanowire array strategy, Biosens. Bioelectron. 25 (2009) 218-223. https://doi.org/ 10.1016/j.bios.2009.06.041

[128] H.-F. Cui, J.-S. Ye, W.-D. Zhang, C.-M. Li, J.H.T. Luong, F.-S. Sheu, Selective and sensitive electrochemical detection of glucose in neutral solution using platinum–lead alloy nanoparticle/carbon nanotube nanocomposites, Anal. Chim. Acta 594 (2007) 175-183. https://doi.org/ 10.1016/j.aca.2007.05.047

[129] X. Luo, A.J. Killard, M.R. Smyth, Reagentless glucose biosensor based on the direct electrochemistry of glucose oxidase on carbon nanotube-modified electrodes, Electroanalysis (N.Y.N.Y.) 18 (2006) 1131-1134. https://doi.org/ 10.1002/elan.200603513

[130] L. García-Gancedo, Z. Zhu, E. Iborra, M. Clement, J. Olivares, A.J. Flewitt, W.I. Milne, G.M. Ashley, J.K. Luo, X. B. Zhao, J.R. Lu, AlN-based BAW resonators with CNT electrodes for gravimetric biosensing, Sens. Actuators B-Chem. 160 (2011) 1386-1393. https://doi.org/ 10.1016/j.snb.2011.09.083

[131] Y. Li, Y.-Y. Song, C. Yang, X.-H. Xia, Hydrogen bubble dynamic template synthesis of porous gold for nonenzymatic electrochemical detection of glucose, Electrochem. Commun. 9 (2007) 981-988. https://doi.org/ 10.1016/j.elecom.2006.11.035

[132] Y.-C. Tsai, S.-C. Li, S.-W. Liao, Electrodeposition of polypyrrole–multiwalled carbon nanotube–glucose oxidase nanobiocomposite film for the detection of glucose, Biosens. Bioelectron. 22 (2006) 495-500. https://doi.org/10.1016/j.bios.2006.06.009

[133] Y.-L. Yao, K.-K. Shiu, Direct electrochemistry of glucose oxidase at carbon nanotube-gold colloid modified electrode with poly (diallyldimethylammonium chloride) coating, Electroanalysis (N.Y.N.Y.) 20 (2008) 1542-1548. https://doi.org/ 10.1002/elan.200804209

[134] S. Park, T.D. Chung, H.C. Kim, Nonenzymatic glucose detection using mesoporous platinum, Anal. Chem. 75 (2003) 3046-3049. https://doi.org/ 10.1021/ac0263465

[135] J. Wang, D.F. Thomas, A. Chen, Nonenzymatic electrochemical glucose sensor based on nanoporous PtPb networks, Anal. Chem. 80 (2008) 997-1004. https://doi.org/ 10.1021/ac701790z

[136] J. Chen, W.-D. Zhang, J.-S. Ye, Nonenzymatic electrochemical glucose sensor based on MnO_2/MWNTs nanocomposite, Electrochem. Commun. 10 (2008) 1268-1271. https://doi.org/ 10.1016/j.elecom.2008.06.022

[137] X. Kang, Z. Mai, X. Zou, P. Cai, J. Mo, A sensitive nonenzymatic glucose sensor in alkaline media with a copper nanocluster/multiwall carbon nanotube-modified glassy carbon electrode, Anal. Biochem. 363 (2007) 143-150. https://doi.org/ 10.1016/j.ab.2007.01.003

[138] M. Zhou, L. Shang, B. Li, L. Huang, S. Dong, Highly ordered mesoporous carbons as electrode material for the construction of electrochemical dehydrogenase-and oxidase-based biosensors, Biosens. Bioelectron. 24 (2008) 442-447. https://doi.org/ 10.1016/j.bios.2008.04.025

Toxic Gas Sensors and Biosensors
Materials Research Foundations **92** (2021) 69-106

Materials Research Forum LLC
https://doi.org/10.21741/9781644901175-3

Chapter 3

2D Materials for Gas and Biosensing Applications

Maneesh Kumar Singh[1], Narendra Pal[2], Sarika Pal[3*], Y.K. Prajapati[4], J.P. Saini[5]

[1]Department of ECE, NIT Uttarakhand Srinagar, Garhwal-246174 (U.K.), India

[2] Department of ECE, Roorkee Institute of Technology, Roorkee- 247667 Uttarakhand, India

[3]Department of ECE, NIT Uttarakhand, Srinagar Garhwal-246174 (U.K.), India

[4]Department of ECE, MNNIT, Allahabad-211004 (U.P.), India

[5]Department of ECE, NSUT, New Delhi-110078, India

*sarikapal@nituk.ac.in, narensarru@gmail.com

Abstract

This chapter discusses the unique and novel properties of 2D materials useful for toxic gas and biosensing applications. The work presented in this chapter mainly focuses on latest research done on 2D materials related to toxic gas and biosensing for surface plasmon resonance based sensors. Here, we proposed a surface plasmon resonance sensor utilizing P3OT thin films which can sense different concentration of NO_2 gas. The performance of proposed design is evaluated by calculating sensitivity, detection accuracy and quality factor, with and without use of silicon layer. Sensitivity of proposed sensor increases by using silicon.

Keywords

Two Dimensional (2D) Materials, Surface Plasmon Resonance (SPR), Gas Sensor, Biosensor, Sensitivity (S)

Contents

Toxic Gas Sensors and Biosensors
Materials Research Forum LLC
Materials Research Foundations **92** (2021) 69-106
https://doi.org/10.21741/9781644901175-3

1.   Introduction to 2D layered materials

Nowadays 2D materials are well suited for sensing applications in the field of healthcare, biomedical, drug-diagnostic, food-safety and environment-safety etc. [1-5]. Use of 2D materials improves sensor performance due to their high surface area to volume ratio and prospects of fabrication in layered form, which gives better response to surface adsorption phenomenon [1,6]. 2D materials are strong entrants for toxic gas and biosensing applications due to their favorable electronic, electrical, mechanical and optical properties [1-4, 6-11]. These unique properties are possible with latest advancement technology for synthesis and engineering of 2D materials. Device integration and scalable manufacturing of 2D material has been possible due to its planar nature and its compatibility with modern fabrication technique [12-13]. 2D material possess unique optical and vibrational properties like photoluminescence, second harmonic generation, fluorescence quenching and enhanced electron–photon interaction useful for optical sensing in addition to tunable electronic and electrochemical properties [14-16].

The use of 2D material in toxic gas and biosensing application requires deep understanding of toxicology of the 2D material i.e. close interaction between 2D layers and sensing analyte. Toxicology of the 2D material is greatly affected by its synthesis methods, material properties, surface chemistry, morphology and its compatibility with interacting layer [17]. Toxicology in the 2D materials affects analyte chemically, electrochemically, mechanically and thus changing compatibility with each other [17-18]. Chemical, electrochemical and mechanical toxicology refers to decomposition of material, catalytic properties of 2D materials and bending or hydrophobicity of the nanosheets respectively [19-22]. So, a thorough understanding to predict toxicology of the 2D material is required before using them in toxic gas or biosensing. Advancement in fabrication technology has made 2D materials much more compatible to gas and biosensing applications.

Nowadays, 2D materials like graphene, black phosphorus (BP), MXene, transition metal oxides (TMOs) and transition metal dichalcogenides (TMDCs) are preferred to enhance the performance of surface plasmon resonance SPR gas and biosensor. These materials do not generate surface plasmon polaritions SPPs to support SPR signal, only their nature of absorption is utilized. Now, unique and fascinating properties of some of the 2D materials useful for SPR gas and biosensor will be discussed here.

1.1   Graphene

It is a single atom layer with sp^2 hybridized carbon atoms having thickness 0.34 nm. It came into existence in 2004 after extraction from graphite. It possesses unique electrical properties (zero bandgap, high charge carrier mobility even at room temperature), morphological properties (high surface area to volume ratio), exceptional mechanical strength and biocompatibility with other materials [11, 21, 23]. Its optical co-absorption efficiency is 2.3% which increases on increasing number of layers [24]. In multilayer graphene based sensor configuration each layer of graphene is bonded by van der Waals forces.

1.2 Black phosphorus (BP)

BP is puckered honeycomb lattice structure of sp^3 hybridized phosphorus atoms which gives it large surface to volume ratio [25]. It possess fascinating electrical properties like high carrier mobility (1000 $cm^2V^{-1}s^{-1}$), tunable bandgap (0.3eV-bulk, 1.9 eV-monolayer BP), layer dependent work function, higher molar response factor, part per billion sensing ability and much higher adsorption energy [25-28]. Like other 2D materials it can be obtained in monolayer form through mechanical exfoliation with scotch tape or by liquid phase epitaxy [29-31].

In multilayer configuration each layer of BP is bonded together through van der Waals (vdw) forces as of graphene. Anisotropic nature of BP, its strong light matter interactions and ultrastrong confinement of SPs suites to design miniaturized optical device [26]. However, oxidation problem in humid nature affects sensor ability [32]. The degradation of BP surface occurs due to materials like metal oxides or polymer films etc. Thus, it can be used efficiently in gas and biosensor due to its extraordinary electronic, morphological properties and ease of fabrication with proper consideration of toxicology of BP in layered form.

1.3 MXene ($Ti_3C_2T_x$)

MXene, a new 2D material have found emerging applications in SPR gas and biosensing [33-34]. Its novel electrochemical properties, extremely high conductivity, fully functionalized and highly accessible hydrophilic surface make it feasible to enhance the performance of the sensor [35-37]. Various theoretical and experimental studies on MXene suggest its use as 2D layered material is because of its novel mechanical, physical, electrical, electronic, optical and plasmonic properties [35].

1.4 Transition metal dichalcogenide (TMDCs)

It possesses chemical composition MX_2, where M denotes transition metal (Tungsten-S, Molybdenum-M) and X denotes chalcogenides (Sulfur-S, Selenium-Se). In these materials, a transition metal atom is bonded to two chalcogen atoms. Thus, using above chemical formula different TMDCs combinations like MoS_2, $MoSe_2$, WS_2, and WSe_2 are possible and can be used in sensing application [38].

Monolayer TMDCs are strongly bonded in plane but in multilayer configuration their interlayers are bonded through weak van der Waals forces. So, monolayer TMDCs can be easily exfoliated from multilayer form. Their unique properties like direct bandgap, high surface area as compared to volume, higher light absorption efficiency, strong light matter interaction make them useful in SPR based gas and biosensor [33, 38-39]. TMDCs can be found in different elemental compositions, synthesis techniques, different structures and layer numbers, but for choosing those in sensing application their toxicology must be considered [40]. Biocompatibility of MoS_2 and WSe_2 was tested by W. Z. Teo and coworkers [41] (2014), for lung epithelial cell, he concluded that WSe_2 exhibits high toxicity than WS_2 and MoS_2 due to toxic effect of Se.

1.5   Transition metal oxides (TMOs)

2D nature and ultrathin thickness of TMOs refers to unique physical, optical, and electronic properties useful for photonic as well as sensing applications [42-43]. Their toxicology must be considered before using them as sensor as they suffer from surface modification, structure decomposition on interaction with biomolecules.

Large amount of research on new 2D material enabled the satisfaction of the increasing demand to develope simple, cost-effective, reliable, portable and highly sensitive sensor. In the past years, different optical sensors have been industrialized which includes surface enhanced Raman spectroscopy (SERS) sensors, surface plasmon resonance (SPR), Forster resonance energy transfer (FRET), as they are immune to electromagnetic interference (EMI) and undesired influence of polarization radiation [14-15, 44]. 2D material based SPR sensors have gained popularity for toxic gas and biosensing due to their superior characteristic of accuracy, reliability, fast, cost-effective, high sensitivity, label-free and real-time sensing competency.

2.   Introduction to SPR sensor

In 1902, Wood [45] discovered phenomenon of SPR which was further explained and experimentally verified by Kretschmann [46] in 1968. In the literature, various types of SPR sensors like prism coupled, waveguide coupled and fiber optic coupled sensors are designed and demonstrated on the basis of light coupling methods [47]. Prism coupled SPR sensors are preferred configuration due to its easier and simple realization. It consists of prism on to which a plasmonic material/metal is deposited conventionally.

Toxic Gas Sensors and Biosensors   Materials Research Forum LLC
Materials Research Foundations **92** (2021) 69-106   https://doi.org/10.21741/9781644901175-3

Different metals like Au, Ag, Al, Cu can be used for surface plasmons (SPs) generation due to richness of free electron in it [48-49]. Au thin film is usually preferred as SPPs metal instead of Ag, Al and Cu as it is stable and cannot be easily oxidized. In Kretschmann configuration based SPR sensor presented in Fig. 1 p-polarized light when coupled through prism, TIR phenomenon occurred at prism/metal interface and evanescent wave is created at metal dielectric interface [46]. On proper matching of wave vector of evanescent field with surface plasmon wave (SPW) field, energy of evanescent wave is completely transferred to SPW and a resonance condition establishes. Thus, on achieving this resonance condition, dip in reflectivity is observed when measured through photodetector. The amount of shift in position of resonance dip is directly related to the extent of variation in sensing layer refractive index. Sensing layer RI shift leads to vary propagation constant of SPs which modifies the resonance condition of SPW and interacting optical wave resulting in shift of resonance dip position. Thus, SPR sensors detect local shift in sensing medium RI in presence or absence of target analyte. SPR signal is much more sensitive to changes in surrounding environment or sensing medium which leads to shifting of dip position of SPR curve due to change in resonance condition.

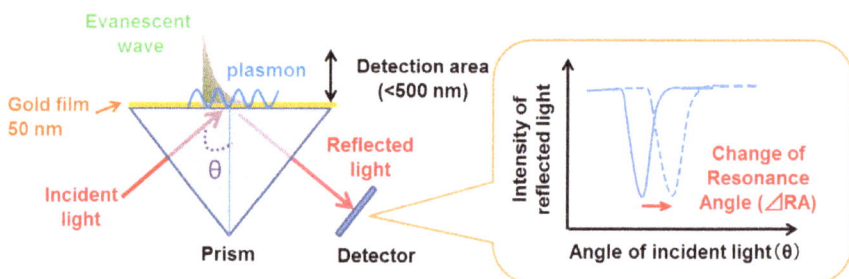

Fig.1 Kretschamann configuration based SPR sensor.

2.1 Theoretical modeling for SPR sensor

The reflectance of a multilayer structure is calculated here through the transfer matrix method (TMM) [50-52]. The transfer matrix method is based on Fresnel's equations. Maxwell's equations can be applied to get expressions for the electric (E) and magnetic (H) fields at each interface in the multilayer structure. It is applied for N-layer model to

Toxic Gas Sensors and Biosensors   Materials Research Forum LLC
Materials Research Foundations **92** (2021) 69-106   https://doi.org/10.21741/9781644901175-3

calculate reflection intensity [50-52]. This method is efficient and free of approximations. But this method only studies homogeneous and isotropic medium.

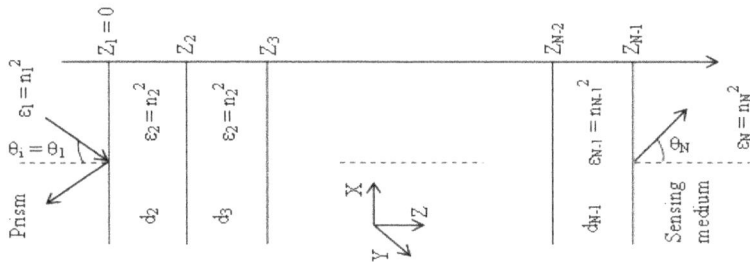

Fig. 2 N-Layer model.

For analysis of Fresnel's equations, we assume it to be smooth and the interfaces between them continuous. The p-polarized light incident at first layer at an angle θ_i and the transmitted light is refracted at angle θ_N. The thicknesses of the layers (d_k), dielectric constant (ε_k) and k^{th} layer RI (n_k) is considered along the z-axis. The tangential fields at $Z=Z1=0$ and $Z=ZN-1$ is calculated as per boundary condition:

$$\begin{bmatrix} U_1 \\ V_1 \end{bmatrix} = F \begin{bmatrix} U_{N-1} \\ V_{N-1} \end{bmatrix} \tag{1}$$

where, $[U_1, V_1]$ and $[U_{N-1}, V_{N-1}]$ denotes the tangential E and H fields at boundary of 1st and N^{th} layer respectively. The F_{ij} denotes characteristics matrix of multilayer structure, which is given as:

$$F_{ij} = \left(\prod_{k=2}^{N-1} F_k \right)_{ij} = \begin{bmatrix} F_{11} & F_{12} \\ F_{21} & F_{22} \end{bmatrix} \tag{2}$$

with,

$$F_k = \begin{bmatrix} \cos \beta_k & (-i \sin \beta_k)/q_k \\ -i q_k \sin \beta_k & \cos \beta_k \end{bmatrix} \tag{3}$$

where,

$$q_k = \left(\frac{\mu_k}{\varepsilon_k} \right)^{1/2} \cos \theta_k = \frac{\left(\varepsilon_k - n_1^2 \sin^2 \theta_1 \right)^{1/2}}{\varepsilon_k} \tag{4}$$

$$\beta_k = \frac{2\pi}{\lambda} n_k \cos \theta_k (z_k - z_{k-1}) = \frac{2\pi d_k}{\lambda} (\varepsilon_k - n_1^2 \sin^2 \theta_1)^{1/2} \tag{5}$$

Using stepwise mathematical procedure, to get reflection intensity coefficient for p-polarized light as:

$$r_p = \frac{(F_{11} + F_{12} q_N) q_1 - (F_{21} + F_{22} q_N)}{(F_{11} + F_{12} q_N) q_1 + (F_{21} + F_{22} q_N)} \tag{6}$$

Reflection intensity, R_p for above described multilayer structure is given as:

$$R_p = |r_p|^2 \tag{7}$$

SPR reflectance curve or reflection intensity curve is the variation of reflection intensity (R_p) corresponding to incidence angle. The reflection intensity becomes least at resonance angle and dip in reflection intensity is obtained [50-52].

2.2 Performance factors for SPR sensor

Major performance factors of the SPR biosensor studied in this chapter are sensitivity, detection accuracy and quality factor and all of them must be high for a quality SPR sensor [50-52].

2.2.2 Sensitivity (S)

It is the primary performance factor that tells us how efficient the sensor is to find RI change of sensing medium. It is the amount of resonance angle shift ($\delta\theta_{Res}$) to deviation in cover or sensing medium RI (δn_c). The resonance angle shift i.e. $\delta\theta_{Res}$ is difference of resonance angle θ_2 and θ_1 as shown in Fig. 3. Where, θ_1 and θ_2 are resonance angles before and after binding of analyte on the sensor surface respectively. Similarly the change in cover or sensing medium refractive index (δn_c) is difference in the values of cover refractive indices of $n_{c2} = 1.41$ to $n_{c1} = 1.33$ as shown in Fig. 3 where n_{c1} and n_{c2} are sensing or cover layer refractive indices before and after the binding of analyte on sensor surface respectively. The more the resonance angle shift, the greater will be the sensitivity. Sensitivity is measured in degree/RIU unit.

$$S = \delta\theta_{Res} / \delta n_c \tag{8}$$

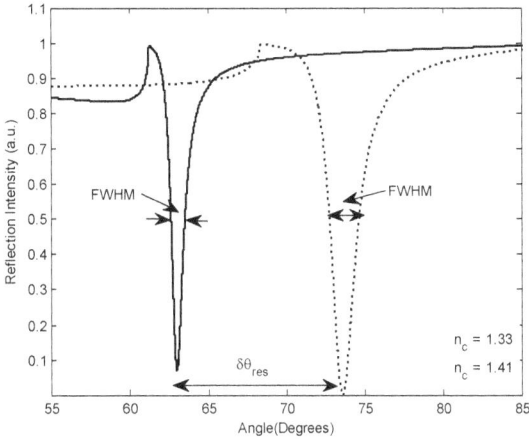

Fig. 3 Reflection intensity curve of SPR sensor.

2.2.2 Detection accuracy (D.A.)

It may also be defined as resolution or signal-to-noise ratio (SNR). FWHM of SPR curve should be narrow, so that there is slightest error in the resolving dip position. The FWHM ($\delta\theta_{0.5}$) is the spectral width of the reflection intensity curve corresponding to 50% reflectivity. It is defined as inverse of FWHM ($\delta\theta_{0.5}$) of the reflection intensity curve. Its unit is degree^{-1}.

$$D.A. = 1/FWHM \tag{9}$$

2.2.3 Quality factor (Q.F.)

It is the product of S and D.A. of the reflectivity curve, It's unit is RIU^{-1}:

$$Q.F. = S*DA \tag{10}$$

2.2.4 Linearity

Reflectivity should vary linearly for wide RI range of sensing region. The range of the refractive indices for which SPR sensitivity varies linearly with variation in refractive indices is known as dynamic range of the SPR sensor.

2.2.5 Reproducibility

It tells us how many times the sensor can be reused. It reduces the cost of sensing related tryouts when the same sensor is able to be used for numerous times.

2.2.6 Stability

It is capability to deliver the same output when assessing same analyte under the same operating conditions over regular intervals. It depends on the robustness of the materials used in the sensor.

3. Application of 2D layered materials for toxic biosensing and gas sensing

Various materials have been investigated from several decades to improve sensor performance [1-4, 50-52]. Sensor performance depends on the materials used, sensor geometries, coupling methods, interrogation mechanism and attachment method of analyte. Different materials used for the sensor substrate, dielectric layers, metal layers and sensing layers deposited for signal transduction based on biomolecular i.e. analyte attachment plays a very imperative role in affecting the performance of the sensor. Much grown research on 2D materials has enabled its use in toxic gas and biosensing applications.

The unique properties of 2D layered materials useful for SPR gas and biosensing has already been discussed in previous section. Now this chapter will discuss the review on using 2D materials for toxic biosensing and gas sensing SPR sensors applications.

3.1 2D Material based SPR biosensor

In developing nations, major public health diseases like typhoid, diarrhea, pneumonia caused by pathogens needs early detection of target analyte [53-56]. So, a highly sensitive and portable biosensor is required for early detection of the target analyte without any delay in intermediate sample processing. SPR biosensors are preferable due to their real-time, label-free, faster and accurate detection methods [47]. Thus, increasing mandate for developing much sensitive, cost effective, reliable, and portable SPR sensor has promoted numerous research for new nanomaterials.

Exponentially growing research on use of 2D material based SPR biosensor for detection of nucleic acids, proteins, bacterial cells, heavy metals, neurotransmitters (serotonin, dopamine) and metabolites (glucose, lactose, adenosine and ascorbic acid) etc. has been observed [1, 57-59]. Thus, this class of material has offered enhanced sensitivity due to their atomically thin layered nature and advanced functionalization and fabrication technologies. This enabled its use in the field of healthcare, disease diagnostic and drug diagnostics etc. 2D materials find their role in diagnostic and vivo application after clear and necessary understanding of their toxicology or biocompatibility [1]. 2D material based sensor can perform early and rapid and label-free detection of nucleic acid in vitro diagnostic [1]. These materials are the basis of next generation technology in personalized medicine with their ability to detect very small concentrations. Now this sub-section will review the work proposed by researchers for 2D materials (graphene, BP, TMDCs, metal oxides and transition metal oxides) based SPR biosensor.

3.1.1 Graphene based SPR biosensor

L. Wu and coworkers [60] (2010), proposed and developed graphene on Au based SPR biosensor giving 25% sensitivity enhancement for 10 layers of graphene. R. Verma and coworkers [61] (2011), further enhanced SPR sensitivity by using high refractive index silicon and graphene layers for sensing of biomolecules at different excitation wavelengths. The sensor's best performance is obtained for optimized thicknesses of Au (40 nm), silicon (7 nm) and two layers of graphene at 633 nm wavelength. Almost double sensitivity is found for optimized sensor than sensors reported in the literature. S. H. Choi and coworkers [62] (2011), presented numerical study that showed graphene based sensor with silver as SPR active metal, to enhance sensing performance of SPR biosensor. It has been found to be superior to conventional gold-based SPR imaging biosensors. P.K. Maharana and R. Jha [63] (2012), proposed a SPR biosensor in visible and NIR region. The sensitivity increases significantly by using graphene, the DA increases further by 100% due to use of chalcogenide prism instead of silica. Also, DA of the proposed sensor in NIR is 16 times more to that in the visible regime.

Kim et al. [64] (2013), experimentally setup fiber optic SPR biosensor using fabricated graphene and replaced the metal films. R. Galatus and coworkers [65] (2013), proposed SPR biosensor, where graphene layers were coated over the multilayer base of the plastic optical fibre (POF). P.K. Maharana and coworkers [66] (2014), proposed an air mediated SPR sensor for sensitivity enhancement in near infrared frequency range. For ten layers of graphene, sensitivity obtained is 43.18°/RIU, and 36.14°/RIU at 700 and 1000 nm operating wavelength respectively. Sensitivity as a function of sensing RI increases for more graphene layers. At known wavelength, multilayer graphene and refractive index of sensing layer improve the sensitivity and DA improved up to 290% compared to gold based SPR sensor in NIR. Verma and coworkers [67] (2015), proposed a graphene based SPR biosensor for detection of bacteria (Pseudomonas). It was perceived that the SPR sensor performance improved with the use of graphene with attachment of bacteria on the sensor surface over the conventional SPR sensor. Verma and coworkers [68] (2015), further enhanced the sensitivity of SPR biosensor for bacteria detection by using air gap along with graphene. The group concluded that the proposed biosensor exhibited sensitivity 2.35 times better than without air gap. Also, it was verified that performance of proposed SPR biosensor get affected through graphene chemical potential while change in temperature does not affect its performance at the same chemical potential.

3.1.2 BP based SPR biosensor

S. Pal and coworkers [69] (2017), proposed two Au-BP based SPR biosensor configuration: first with a silicon layer that increases the detection accuracy but decreases the sensitivity of SPR and the second without silicon. To enhance biosensor performance, thickness of Au layer has been optimized in the visible range. Unique properties of BP like tunable direct bandgap, tunable work function, extraordinary charge carrier mobility, and superior binding strategy of molecules on the sensor surface due to its much higher surface area to volume ratio are utilized. The sensitivity of proposed sensor without silicon is much better as compared to a graphene based and conventional SPR biosensor. Sensitivity of SPR biosensor is 1.42, 1.40 times of conventional and graphene-based SPR biosensor respectively with the use of BP [69].

DNA hybridization is the biochemical method to recognize the unknown nucleotide sequence (target DNA) with known nucleotide sequence (probe DNA) [70]. S. Pal et al. [71] (2018), designed a graphene deposited BP SPR biosensor to detect DNA hybridization. It can be used to diagnose life threating diseases such as cancer and hepatitis B. Here ds-DNA hybridization is detected by hybridization of ssDNA (targets) onto surface immobilized DNA probes. Use of graphene provides the strong and stable π stacking force between target and analyte, which results in changing the refractive index which in turn leads to a change in reflectivity monitored by the SPR sensor. Highest sensitivity of (125°/ RIU), D.A (0.95) and Q.F (13.62 RIU^{-1}) is obtained for proposed BP and graphene based SPR biosensor [71]. Here BP offer excellent absorbability of light and has much faster response in comparison to other transition metal di-chalcogenide materials (TMDC). Meshginqalam B. and J. Barvestani [72] used silicon and BP based SPR biosensor which gives high sensitivity equal to 342°/RIU with 5 layer of BP material. Bimetallic (Al/Au) structure is used as novel SPR active metal to protect Al from oxidation as well as to enhance the sensitivity.

3.1.3 TMDC based SPR biosensor

S. Zenga et al. [73] (2015), proposed a SPR configuration of thin film Au, coating 3 layers of MoS_2 nanosheets and monolayer graphene hybrid structure has obtained excellent detection sensitivity. Enhancement factor of phase sensitivity of graphene - MoS_2 hybrid structure based SPR is 500 fold times more than only graphene coating SPR. The monolayer of graphene and monolayer MoS_2 provide strong π stacking force between target and analyte to detect the target, and higher optical adsorption efficiency respectively. Maurya and coworkers [24] (2016), presented a SPR biosensor based on molybdenum disulfide. 2D material is used to detect Pseudomonas like bacteria which cause severe acute and chronic infections in respiratory and urinary tract and wounded skin. Monolayer and bilayer MoS_2 is used in the proposed SPR biosensor, which has advantages like high adsorption efficiency and tunable band gap. Proposed SPR biosensor enhances the performance parameter to detect the Pseudomonas like bacteria to graphene based and the conventional SPR biosensor.

N.A. Jamil et al. [74] (2017), used Au-MoS_2-Graphene based SPR sensor to detect urea with maximum sensitivity 230 °/RIU and 173°/RIU at 670 and 785 nm respectively . Group compared the sensitivity of advanced SPR sensor with and without graphene. SPR biochemical sensor designed by L. Wu et al. [75] (2017), utilizes BP and TMDCs (MoS_2, $MoSe_2$, WS_2, WSe_2) layers to get higher sensitivity. Highest sensitivity is obtained for two layers of WS_2 at 633 nm operating wavelength. Q. Ouyang et al. [39] (2016), theoretically investigated TMDC based SPR biosensor (Au/Si/TMDC) to enhance sensitivity at different operating wavelengths. Highest sensitivity, 155.68 ° /RIU was obtained for Au (35 nm)/Si (7 nm)/ WS_2 (Monolayer) at 600 nm wavelength. S. Pal and coworkers [76] (2019), enhanced the sensor ability to detect biomolecules by using Si-BP-TMDC coated SPR biosensor. Sensitivity (184.6°/RIU) for SPR biosensor at the Au (50 nm), Si (7 nm) and BP (0.53 nm) is attained without using TDMs layers due to exceptional detecting capability of BP and higher field intensity at the Au interface due to the high RI of silicon. The group opened a way to use heterostructure of BP-TMDC in SPR biosensors. Meshginqalam B. and J. Barvestani [77] (2018), developed SPR biosensor using the 10 layer BP and mono layer TMDC (WS_2) over an Au layer to enhance the sensitivity up to 187 °/RIU. They compared the performance of proposed SPR biosensor with other 2D TMDC material based SPR biosensor too.

3.1.4 MXene based SPR biosensor

L. Wu et al. [34] (2018), demonstrated that layered MXene and different SPR active metal based SPR biosensor, improved the sensitivity from 16.8% to 46.3% due to their absorption properties with large surface area and as protective layer top on metal layer,

Materials Research Forum LLC
https://doi.org/10.21741/9781644901175-3

protecting metal from oxidization. Y. Xu et al. [33] (2019), proposed highly sensitive SPR biosensor by utilizing unique properties of MXene and TMDCs. Highest sensitivity (198°/RIU), with sensitivity enrichment of 41.43% was accomplished in aqueous solutions for monolayer MXene ($Ti_3C_2T_x$) and 5 layers of WS_2 at a 633 nm excitation wavelength. A. Srivastava and coworkers [78] (2019), theoretically studied an innovative SPR sensor based on prism-gold layer-MXene-WS_2-BP to detect impure water. The group compared proposed SPR sensor performance with and without MXene. Even though all configuration can detect impure water at 633 nm wavelength but due to excellent electrical, optical and mechanical properties of MXene, sensitivity of MXene and TMDC based SPR has highest sensitivity of 190.22°/RIU.

3.1.5 Metal oxide / transition metal oxides based SPR biosensor

N.A.S. Omar et al. [79] (2019), reported on Au-CdS quantum dot-reduce graphene oxide-antibodies film used to quantitative detection of dengue virus. Dengue viral fever is a mosquito-borne disease, the most dangerous diseases found in the world, transmitted by the female Aedes mosquito, Aedes aegypti. N.A.S. Omar et al. [79] (2019), also reported the chitosan graphene oxide/cadmium sulfide quantum dots based SPR biosensor in the same paper for cobalt ion (Co^{2+}) detection which is a poisonous heavy metal and leads to serious health problems. River, forest, lake, soils and sea are polluted by excessive heavy metal ions which are released by industries, thus affect the natural cycle. G. Kaur and coworkers [80] (2016), realized an efficient SPR based DNA biosensor for detection of Neisseria Menningitidis DNA by using a ZnO thin film. The proposed biosensor demonstrated high specificity and prolonged shelf life.

Very recently, S. Pal and coworkers [81] (2019), proposed a ZnO and graphene based SPR biosensor which has been numerically analyzed DNA hybridization for different chemical potential of graphene at 25 °C. The highest sensitivity of the SPR biosensor for DNA hybridization is achieved at 1.25eV chemical potential of graphene.

3.2 2D material based SPR gas sensor

Over the past few decades the standards of living of the human race has risen at phenomenal rate due to the industrialization revolution. Extension of industrialization and chemical technologies causes the discharge of toxic matter in numerous forms, their effect impact on our health and environment. In this growing era, food and water safety and environment pollution monitoring have become priority area for society safety. Air pollution contains harmful chemicals or particles in it, which harm humans, animals and the overall plant health. In air pollution a large number of toxic gases and toxic particles are present that can lead to many health problems such as long term disease and metal poisoning. Such toxic gases have the capability to harm the living tissues, impairment of the central nervous system, result in illness or in extreme cases death when such type of gas is inhaled or absorbed by eyes and skin. Some poisonous gases are not visible and cannot be smell, and they will not have an immediate effect but may be the cause of death. Toxic gases including hydrogen sulphide (H_2S), carbon monoxide (CO), ammonia (NH_3), nitrogen dioxide (NO_2), formaldehyde, methyl are produced from various area such as industries (petroleum, chemical, sugar, coal, gas, mines), sewage sludge, waste disposal, vehicle and natural disaster (forest fire) [82-83].

It is essential that quick and precise detection, monitoring and alerting of toxic gases is done to minimize accidents involving poisoning or explosions. Most of the researchers and known research centers have focused their interest on designing of effective gas sensors and on investigations of sensor material to enhance the sensitivity of gas sensors. There are varieties of sensing procedures which rely on different chemical or physical phenomena. The presence of target analytes can be detected by change in electrical properties (resistance, conductivity, capacitance), optical properties (transmission, absorption, reflection, refractive index, luminescence), body properties (thermal conductivity, temperature, mass, propagation of acoustic waves) analyzing electrochemical or biochemical recognition etc. [82-87]. In the literature, various types of methods for detection of toxic gas have been reviewed. These are based on potentiometric electrodes, variation of metal oxide conductivity, dye absorption, surface acoustic wave etc. [82-87]. Such sensors have many drawbacks, such as higher costs, poor reliability, smaller sensitivity, non-reusability, high operating temperature etc. Recent years have seen 2D materials significantly used in SPR sensors for detection of toxic gases [4-5, 11]. Some of the 2D material based SPR sensors for toxic gas will be discussed here.

3.2.1 Graphene based SPR gas sensor

J. Yang et al. [88] (2015), designed a long period fiber grating (LPFG) based methane gas sensor by surface deposition of high RI polycarbonate/Cryptophane A. Maximum sensitivity of 3.56×10^3 nm/RIU is obtained for surface thickness of 530 nm, RI of 1.5605 and grating period of 520 μm. W. Wei et al. [89] (2017), proposed highly sensitive graphene based LPFG SPR sensor for gas sensing. Here monolayer graphene is coated onto a silver layer of LPFG to increase the magnitude of the evanescent field on the core surface of the fiber resulting in enhanced interaction of the SPR wave with the gas molecules. The group experimentally verified that maximum sensitivity achieved could be of 0.344 nm $\%^{-1}$, which is 2.96 times to conventional and 1.31 times to silver coated SPR.

T. Srivastava and coworkers [90] (2016), proposed a graphene based SPR gas sensor in the terahertz region for modified Otto geometry using attenuated total reflection (ATR) geometry [90]. In this geometry, an airgap is replaced by a dielectric spacer of organic material having RI 1.44, 1.50 and 1.54 at frequency 5 THz and performance of the sensor is evaluated through sensitivity, DA and FoM. It is perceived that proposed SPR gas sensor exhibits a tradeoff between sensitivity and detection accuracy. However, obtained figure of merit is 20% higher at organic material refractive index (RI) of 1.54 in comparison to RI of 1.44 and 1.50 of organic material. A. Purkayastha and coworkers [91] (2016), proposed ultrasensitive terahertz plasmonic SPR gas sensor using doped graphene. He evaluated performance of proposed Otto configuration based SPR sensor through ATR technique. Sensitivity and FoM were evaluated at different chemical potential of monolayer graphene for proposed design. Proposed sensor shows sensitivity, 34.11°/RIU along with ultrahigh FoM greater than 1150 RIU^{-1}. P.K. Maharana and coworkers [92] (2015), proposed graphene on Ag based high performance affinity gas sensor operating in visible and near infrared region. Imaging sensitivity obtained for proposed sensor was 440, 150 and 100% higher to silicon on Ag based SPR design at operating wavelengths 653, 850 and 1000 nm respectively. Performance is improved due to the higher electric field enhancement on use of graphene.

3.2.2 BP based SPR gas sensor

Maurya and coworkers [93] (2018), proposed a Si-BP based SPR sensor for NO_2 gas sensing. The proposed design utilizes attachment method of NO_2 gas on BP surface. Silicon is used as high RI layer for sensitivity enhancement. L Wu et al. [94] (2018), investigated a few layers BP based high performance lossy mode resonance sensor. Proposed sensor obtained maximum Q factor of $2x10^5$ RIU^{-1} for TM polarized light. Y. Singh and coworkers [95] (2019), enhanced the sensitivity of the BP based SPR gas sensor by 80.5 % in comparison to conventional SPR sensor. Sensitivity is further increased to 102.39% on adding extra BP layers.

3.2.3 TMDCs based SPR gas sensor

H. Wang et al. [96] (2018), proposed a sensitivity enhanced SPR gas sensor by using a WS_2 nanosheet overlayer. Peak sensitivity of 2459.3nm/RIU is attained for proposed design due to high RI, larger surface area and unique optoelectronic properties of WS_2 nanosheet. G.A Asres et al. [97] (2018), designed a p-type WS_2 hybrid material based ultrasensitive H_2S gas sensor. Their work demonstrated excellent sensitivity of and high selectivity to H_2S relative to NO, CO, H_2 and NH_3 gas sensing. Pristine WS_2 was not able to give high sensitivity so O doping in the S sites of WS_2 lattice was used to display ultrahigh sensitive behavior. A. Sharma and A. Pandey [98] (2018), investigated heterostructure of blueP/ MoS$_2$ based highly sensitive SPR sensor for gas and biosensing in visible and infrared region. Proposed design exhibits highest sensitivity (432.15°/RIU) at 662 nm operating wavelength.

3.2.4 MXene (Ti$_3$C$_2$T$_x$) based SPR gas sensor

E. Lee et al. [99] (2017), performed ammonia gas sensing by using 2D MXene (Ti$_3$C$_2$T$_x$) at room temperature. The MXene nanosheet was synthesized after elimination of Al atoms from MAX phases of MXene and was fabricated on polyimide platforms with a simple solution based process. Proposed sensor also measured methanol, ethanol, acetone gas at room temperature and limit of detection obtained for acetone gas was 9.27 ppm after theoretical calculation. Kim et al. [36] (2018), utilized high metallic conductivity and fully functionalized surface of MXene to design a gas sensor. L. Lorencova et al. [100] (2017), designed a MXene based electrochemical sensor for hydrogen peroxide (H$_2$O$_2$) sensing.

3.2.5 Metal oxides / transition metal oxides based SPR gas sensor

S.K. Mishra and coworkers [101] (2012), designed an ammonia gas sensor based on fiber optic SPR principle for wavelength interrogation as well as intensity modulation technique. The performance of the gas sensor by using ITO (indium tin oxide) as a SPR active material is much better than using gold instead. Resonance wavelength shift is due to reactivity of ammonia with polyaniline which act as dialectical on ITO.

B.D. Gupta and coworkers [102] (2013), experimentally studied on ZnO based (SPR) for detection of toxic gas H$_2$S with sensing range of 10 ppm to 100 ppm. They compared developed SPR gas sensor to other metal oxide (TiO$_2$, SiO$_2$, SnO$_2$). Proposed ZnO based gas sensor has highest sensitivity due to use of ZnO has large active surface area compare to other metal oxide which is tested by photoluminescence spectra. Electrical, chemical and optical properties of ZnO metal oxide are changed when H$_2$S gas interacts with ZnO. They compared ZnO based SPR sensor with chemical sensors based on ZnO and CuO with advantages of operated at room temperature, high surface to volume ratio and response and recovery time approximately 1 minute. A. Paliwal and coworkers [103] (2017), developed a SPR sensor using ZnO as a sensing material to detect carbon mono oxide (CO) toxic gas. They optimized 200 nm thick layer of ZnO grown on gold layer at 250 °C exhibits high sensitivity. They also show the selectivity of the gas sensor in terms of reflectance towards CO with other gases (NH$_3$, CO$_2$, NO$_2$, LPG, H$_2$S) which is almost negligible.

Materials Research Forum LLC

https://doi.org/10.21741/9781644901175-3

4. Proposed SPR sensor based on doped and undoped polymer (P3OT) thin film for NO₂ gas detection

Detection of NO_2 gas for environment monitoring and industrial control using inorganic semiconductors like WO_3, SnO_2 and In_2O_3 has been studied widely in the past few years [104-106]. This involves high operating temperatures (150-450 °C) and low detection range of about 10-100 ppm [107]. The use of polymer instead of inorganic material offers many advantages like wide choice of molecular structure, lower cost materials, and easier fabrication techniques which enables thin film possessing chemical properties useful for sensing at room temperature [108]. Some examples of semiconductor polymers studied in the past are polyaniline, polypyrrole, polyenylenevinylene, and polythiophene [109-110]. Among these polymers especially polythiophene has been used for detection of different gasses. J.C. Solis and coworkers [111] (2006), performed optical characterization of spin-coated P3OT thin films beneath NO_2 gas exposure at 5, 10, 3000 ppm [111].

P3OT thin films developed here consists of conjugated double bonds along its mainstay, with alternate single and double bonds in between carbon atoms. The π bond (double bond) is stronger compare to sigma bond (single bond). This conjugated structure of double bonds shows oxidative doping effects to NO_2 gas making polymer conductive. Refractive index of these thin films changes under NO_2 doping effect may be utilized for SPR based NO_2 gas sensing application. Here in this chapter, we propose a P3OT thin film based SPR gas sensor for NO_2 gas sensing.

4.1 Proposed sensor design

Proposed sensor design is based on N-layer modeling discussed in previous section for calculating reflection intensity of the SPR sensor. Proposed SPR gas sensor is shown in Fig. 4. A CaF_2 prism is used here, having RI of 1.4329 at operating wavelength of 633 nm. A silver layer of 50 nm thickness is laid onto the flat face of the prism. RI of a silver metal layer is taken as per the Drude model. A high RI (3.916 RIU) of silicon layer having 5 nm thickness is grown over an Ag layer to protect it from oxidation.

The second and last layer of P3OT thin film having 35nm thickness is deposited over the silicon layer for NO_2 gas sensing. RI for doped and undoped thin films of P3OT is shown in Table 1 [111]. The last layer consists of the sensing medium (RI, 1-1.1) to detect NO_2 gas absorption on the P3OT thin films [93].

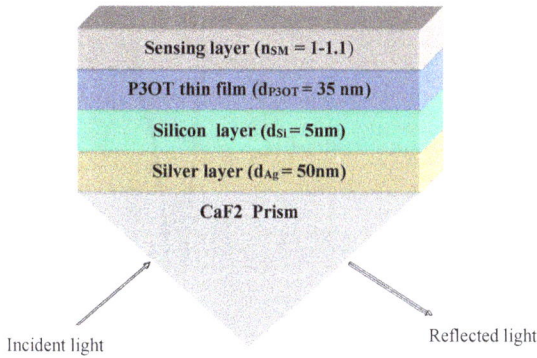

Fig. 4 Proposed SPR sensor design.

Table 1 Values of refractive indices of undoped & doped P3OT thin films [111].

S. No.		Undoped-P3OT thin film	Doped- P3OT thin films		
			5 ppm NO₂	10 ppm NO₂	3000 ppm NO₂
1.	Refractive index values	1.7099	1.6119	1.5927	1.5292

4.2   Results and discussion for proposed sensor design

As per proposed sensor modeling, reflectivity curves are plotted here to evaluate the performance of proposed gas sensor design. Figs. 5a-b show reflectivity curves for undoped P3OT thin film with and without silicon. Resonance angle and minimum reflectance obtained from Fig. 5a for undoped P3OT thin film without silicon is [54.808°, 0.0017a.u.] and [64.049°, 0.0063a.u.] at sensing medium with refractive index (RI) of 1 and 1.1 respectively.

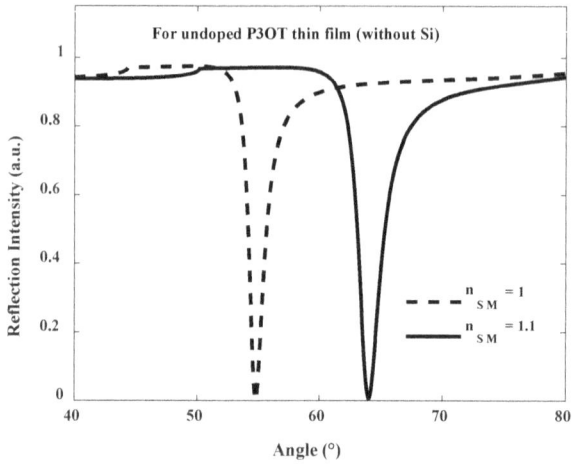

(a)

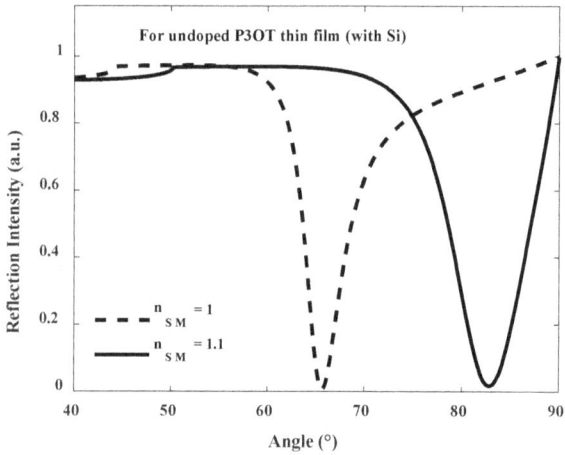

(b)

Fig. 5 Reflection intensity curve of proposed sensor design for undoped P3OT thin film a. Without Si layer b. With Si layer.

Similarly, Fig 5b indicates resonance angle and minimum reflectance obtained for undoped P3OT thin film with silicon is [65.699 °, 0.0071a.u.] and [82.822 °, 0.0182a.u.] at sensing medium RI 1 and 1.1 respectively. Thus, change in resonance angle, sensitivity, DA and quality factor (Q.F.) calculated for undoped P3OT thin film without silicon and with silicon is [9.241°, 92.41° /RIU, 0.657 Degree^{-1}, 60.8 RIU^{-1}], and [17.12°, 171.2° /RIU, 0.199 Degree^{-1}, 34 RIU^{-1}] respectively. Thus, the results indicate that the use of a high RI silicon layer enhances sensitivity by increasing NO_2 absorption on thin films.

Similarly, Figs. 6, 7 and 8 show reflectivity curve for doped P3OT thin films of 5 ppm, 10 ppm and 3000 ppm of NO_2 with and without Si layer respectively. Variation in resonance angle, sensitivity, DA and Q.F. determined from Figs. 6, 7 and 8 with and without using silicon layer is given in Table 2. It can be evaluated from Table 2 that sensitivity decreases for higher concentration of NO_2 due to decreasing trend of RI of thin P3OT thin films for higher concentrations of NO_2 adsorption on it. It can also be concluded from table 2 that the use of a Si layer enhances the sensitivity for the proposed SPR gas sensor. D.A. and Q.F. reduces with the use of a silicon layer due to widening of reflectivity curves.

Table 2 Comparative table of performance parameters obtained for proposed NO_2 gas sensor with and without silicon.

S. No.	Type of P3OT thin film	Change in resonance angle[Degree]		Sensitivity (S) [Degree/RIU]		Detection accuracy (D.A.) [Degree^{-1}]		Quality factor (Q.F.) [RIU^{-1}]	
		Without Silicon	With Silicon	Without Silicon	With Silicon	Without Silicon	With Silicon	Without Silicon	With Silicon
1.	Undoped thin film	9.241	17.12	92.4	171.2	0.657	0.199	60.8	34
2.	Doped thin film- 5 ppm NO_2	8.409	12.45	84.1	124.6	0.820	0.278	68.93	34.6
3.	Doped thin film- 10 ppm NO_2	8.265	11.93	82.7	119.3	0.855	0.295	70.7	35.19
4.	Doped thin film- 3000 ppm NO_2	7.83	10.56	78.3	105.6	0.98	0.356	76.76	37.6

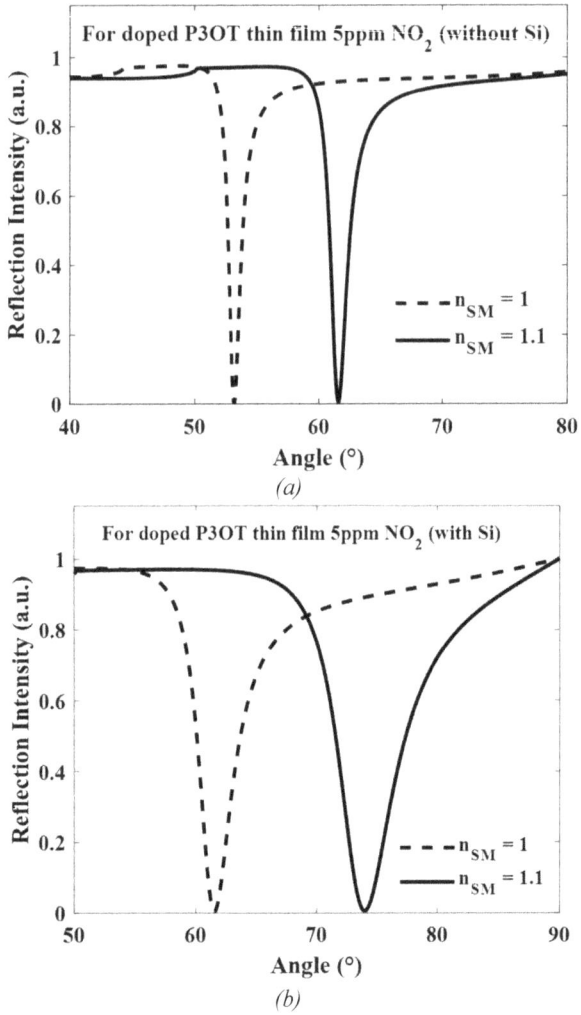

Fig. 6 Reflection intensity curve for proposed sensor design for doped P3OT thin film with 5ppm NO₂ gas - a. Without Si layer b. With Si layer.

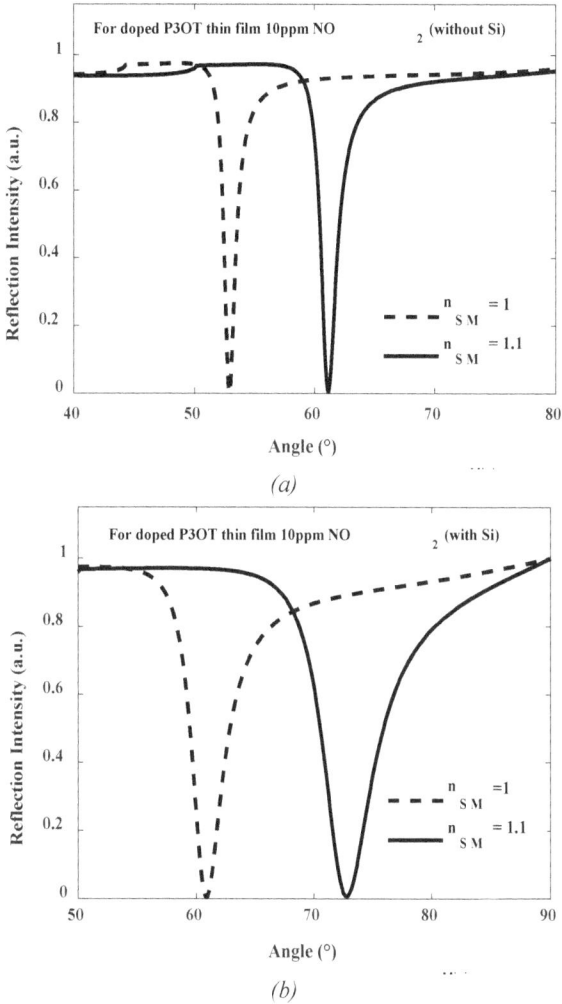

Fig. 7 Reflection intensity curve for proposed sensor design for doped P3OT thin film with 10 ppm NO₂ gas - a. Without Si layer b. With Si layer

(a)

(b)

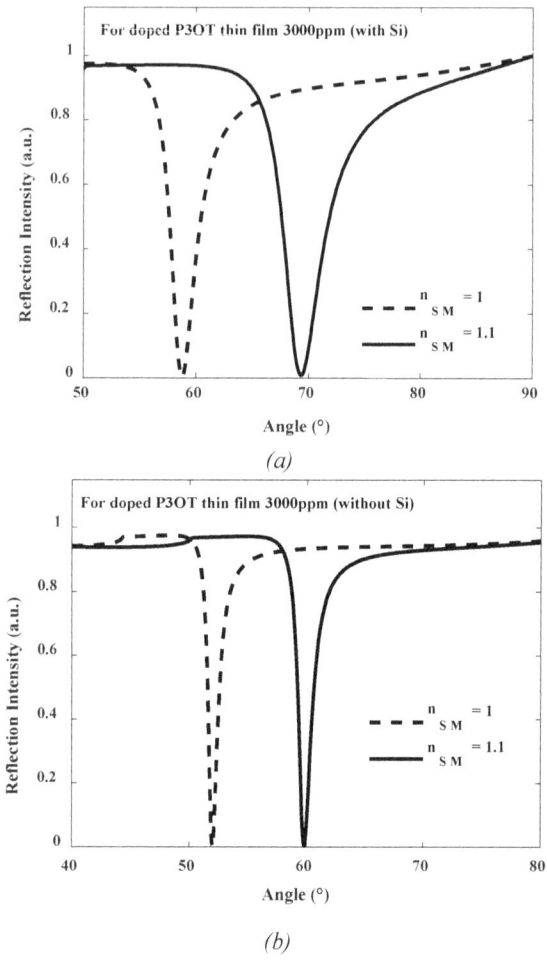

Fig. 8 Reflection intensity curve for proposed sensor design for doped P3OT thin film with 3000ppm NO₂ gas - a. Without Si layer b. With Si layer.

Lastly, Fig. 9a-b is plotted for proposed sensor design to indicate change in resonance angle with respect to P3OT thin film thickness variation from 5 nm to 40 nm for undoped and doped P3OT thin films with and without silicon layer. Fig. 9a shows change in resonance angle increases from 6.99° to 9.843° for undoped P3OT thin film thickness variation of 5-40 nm, without using silicon layer for proposed sensor design. This means sensitivity of proposed sensor increases on increasing P3OT thin film thickness due to structural change in P3OT thin films on interaction with NO_2 gas. Fig. 9a indicates same increasing trend for change in resonance angle from [6.947°, 6.938°, 6.909°] to [8.687°, 8.5°, 7.946°] for higher thicknesses of doped P3OT thin films with 5 ppm, 10 ppm, 3000 ppm of NO_2 concentration for proposed sensor design without silicon.

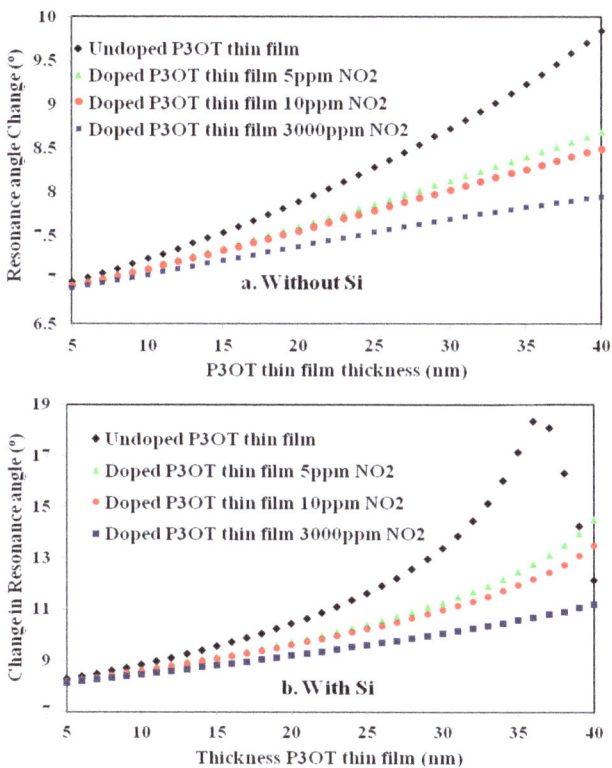

Fig. 9 Curve for change in resonance angle vs. P3OT thin film thickness for undoped and doped P3OT thin film of proposed sensor - a. Without Silicon b. With Silicon.

Fig. 9b indicates change in resonance angle obtained for different thicknesses of doped and undoped P3OT thin films (5 nm-40 nm) for proposed SPR gas sensor on using silicon layer. Change in resonance angle increases from [8.288°, 8.2°, 8.182°, 8.122°] to [12.142°, 14.531°, 13.468°, 11.191°] for undoped, doped P3OT thin film with 5 ppm, 10 ppm, 3000 ppm NO_2 gas respectively, for proposed SPR based NO_2 gas sensor with silicon. On comparing both Figs 9a and 9b, it is observed that use of silicon layer in proposed sensor increases sensitivity. Thus, it may be concluded that proposed sensor can efficiently detect different concentrations of NO_2, a toxic gas harmful to environment and industry.

Conclusions

Here in this chapter, unique and novel properties of different 2D materials applicable to sensing applications is discussed. Then the role of different 2D materials like graphene, BP, TMDCs, MXene, etc. in SPR gas and biosensing application is reviewed. This chapter reviewed most of the research work performed recently in the field of 2D nanomaterial based SPR sensors for gas and biosensing. Here, the mathematical modeling i.e. transfer matrix method used for calculation of reflection intensity is also presented. Thereafter, we proposed P3OT thin film based SPR sensors for NO_2 gas sensing. The proposed sensors performance are evaluated with and without the use of silicon through reflectance curves The performance of proposed gas sensor with and without silicon for doped and undoped P3OT thin films is evaluated in terms of sensitivity, detection accuracy and quality factor. Sensitivity of proposed sensor increases with the use of silicon due to incrementing electromagnetic field at the sensing layer interface. Proposed SPR sensors can efficiently sense different concentrations of NO_2.

References

[1] A. Bolotsky, D. Butler, C. Dong, K. Gerace, N. R. Glavin, C. Muratore, J. A. Robinson, A. Ebrahimi, Two-Dimensional materials in biosensing and healthcare: From in vitro diagnostics to optogenetics and beyond, ACS Nano 13 (2019) 9781-9810. https://doi.org/10.1021/acsnano.9b03632

[2] G. Maduraiveeran, M. Sasidharan, V. Ganesan, Electrochemical sensor and biosensor platforms based on advanced nanomaterials for biological and biomedical applications, Biosen. and Bioelectron., 103 (2018) 113-129. https://doi.org/10.1016/j.bios.2017.12.031

[3] B. D. Malhotra & S. Srivastava, M. A. Ali, C. Singh, Nanomaterial-based biosensors for food toxin detection, Appl. Biochem. Biotechnol. 174 (2014) 880–896. https://doi.org/10.1007/s12010-014-0993-0

[4] S. Yang, C. Jiang, S. H. Wei, Gas sensing in 2D materials, Applied Physics Reviews. 4 (2017) 021304. https://doi.org/10.1063/1.4983310

[5] K. Shehzad, T. Shi, A. Qadir, X. Wan, H. Guo, A. Ali, W. Xuan, H. Xu, Z. Gu, X. Peng, J. Xie, L. Sun, Q. He, Z. Xu, C.o Gao, Y. Rim, Y. Dan, T. Hasan, P. Tan, E. Li, W. Yin, Z. Cheng, B. Yu, Y. Xu, J. Luo,X. Duan, Designing an efficient multimode environmental sensor based on graphene–silicon heterojunction, Adv. Mater. Technol. 2 (2017) 1600262. https://doi.org/10.1002/admt.201600262

[6] R. Kurapati, K. Kostarelos, M. Prato, A. Bianco, Biomedical uses for 2D materials beyond graphene: Current advances and challenges ahead, Adv. Mater. 28 (2016) 6052−6074. https://doi.org/10.1002/adma.201506306

[7] X. Gan, H. Zhao, X. Quan, Two-Dimensional MoS_2: A Promising building block for biosensors, Biosens. Bioelectron. 89 (2017) 56−71. https://doi.org/ 10.1016/j.bios.2016.03.042

[8] K. Shavanova, Y. Bakakina, I. Burkova, I. Shtepliuk, R. Viter, A. Ubelis, V. Beni, N. Starodub, R. Yakimova, V. Khranovskyy, Application of 2D non-graphene materials and 2D oxide nanostructures for biosensing technology, Sensors. 16 (2016) 223. https://doi.org/10.3390/s16020223

[9] M. Xu, T. Liang, M. Shi, H. Chen, Graphene-like two-dimensional materials, Chem. Rev. 113 (2013) 3766−3798. https://doi.org/10.1021/cr300263a

[10] M. Holzinger, A. L. Goff, S. Cosnier, Nanomaterials for biosensing applications: A review, Front. Chem. 2 (2014) 63. https://doi.org/10.3389/fchem.2014.00063

[11] S. Varghese, K. Singh, S. Swaminathan, S. Varghese, V. Mittal, Two-dimensional materials for sensing: graphene and beyond, Electronics. 4 (2015) 651−687. https://doi.org/10.3390/electronics4030651

[12] Z. Lin, B. R. Carvalho, E. Kahn, R. Lv, R. Rao, H. Terrones, M. A. Pimenta, M. Terrones, Defect engineering of two-dimensional transition metal dichalcogenides, 2D Mater. 3 (2016) 022002. https://doi.org/10.1088/2053-1583/3/2/022002

[13] S. Zhang, R. Geryak, J. Geldmeier, S. Kim, V. V. Tsukruk, Synthesis, assembly, and applications of hybrid nanostructures for biosensing, Chem Rev. 117 (2017) 12942−13038. https://doi.org/10.1021/acs.chemrev.7b00088

[14] N. Kamaruddin, A. A. Bakar, N. Mobarak, M. S. Zan, N. Arsad, Binding affinity of a highly sensitive Au/Ag/Au/chitosan-graphene oxide sensor based on direct

detection of Pb^{2+} and Hg^{2+} ions, Sensors (Basel). 17 (2017) 2277. https://doi.org/10.3390/s17102277

[15]　M. Shorie, V. Kumar, H. Kaur, K. Singh, V.K. Tomer, P. Sabherwal, Plasmonic DNA hotspots made from tungsten disulfide nanosheets and gold nanoparticles for ultrasensitive aptamer-based SERS detection of myoglobin, Mikrochim. Acta. 185 (2018) 158. https://doi.org/10.1007/s00604-018-2705-x

[16]　A. A. Ebrahimi, K. Zhang, C. Dong, D. Butler, A. Bolotsky, Y. Cheng, J.A. Robinson, FeSx-graphene heterostructures: Nanofabrication-compatible catalysts for ultra-sensitive electrochemical detection of hydrogen peroxide, Sensors Actuators B. Chem. 285 (2019) 631–638. https://doi.org/10.1016/j.snb.2018.12.033

[17]　Z. Wang, W. Zhu, Y. Qiu, X. Yi, A. V. D. Bussche, A. Kane, H. Gao, K. Koski, R. Hurt, Biological and environmental interactions of emerging two-dimensional nanomaterials, Chem. Soc. Rev. 45 (2016) 1750−1780. https://doi.org/10.1039/c5cs00914f

[18]　V. C. Sanchez, A. Jachak, R. H. Hurt, A. B. Kane, Biological interactions of graphene-family nanomaterials: An interdisciplinary review, Chem Res Toxicol. 25 (2012) 15−34. https://doi.org/10.1021/tx200339h

[19]　A. M. Pinto, I. C. Gonçalves, F. D. Magalhães, Graphene-based materials biocompatibility: A review, Colloids Surf. B. 111 (2013) 188−202. https://doi.org/10.1016/j.colsurfb.2013.05.022

[20]　X. Guo, N. Mei, Assessment of the toxic potential of graphene family nanomaterials, J. Food Drug Anal. 22 (2014) 105−115. https://doi.org/10.1016/j.jfda.2014.01.009

[21]　E. L. K. Chng, M. Pumera, Toxicity of graphene related materials and transition metal dichalcogenides, RSC Adv. 5 (2015) 3074−3080. https://doi.org/10.1039/C4RA12624F

[22]　L. M. Guiney, X. Wang, T. Xia, A. E. Nel, M. C. Hersam, Assessing and mitigating the hazard potential of two-dimensional materials, ACS Nano. 12 (2018) 6360−6377. https://doi.org/10.1021/acsnano.8b02491

[23]　A. K. Mishra, A. K. Mishra, R. K. Verma, Graphene and beyond graphene MoS_2: a new window in surface-plasmon-resonance based fiber optic sensing. The Journal of

Physical Chemistry C, 120 (2016) 2893–2900. https://doi.org/ 10.1021/acs.jpcc.5b08955

[24]   J. B. Maurya, Y. K. Prajapati, R. Tripathi, Effect of molybdenum disulfide layer on surface plasmon resonance biosensor for the detection of bacteria, Silicon. 10 (2018) 245–256. https://doi.org/10.1007/s12633-016-9431-y

[25]   P. Bridgman, Two new modifications of phosphorus, J. Am. Chem. Soc. 36 (1914) 1344-1363 https://doi.org/10.1021/ja02184a002

[26]   F. Xia, H. Wang, Y. Jia, Rediscovering black phosphorus as an anisotropic layered material for optoelectronics and electronics, Nat Commun. **5** (2014) 4458. https://doi.org/10.1038/ncomms5458

[27]   Y. Cai, G. Zhang, Y. W. Zhang, Layer-dependent band alignment and work function of few-layer phosphorene, Sci. Rep. 4 (2014) 6677–6682. https://doi.org/ 10.1038/srep06677

[28]   S. Y. Cho, Y. Lee, H. J. Koh, H. Jung, J. S. Kim, H. W. Yoo, J. Kim, H. T. Jung, Superior chemical sensing performance of black phosphorus: comparison with MoS_2 and graphene, Adv. Mater. 28 (2016) 7020–7028. https://doi.org/ 10.1002/adma.201601167

[29]   L. Kou, Phosphorene: fabrication, properties, and applications, J. Phys. Chem. Lett. 6 (2015) 2794-2805. https://doi.org/10.1021/acs.jpclett.5b01094

[30]   H. Liu, A. T. Neal, Z. Zhu, Z. Luo, X. Xu, D. Tomanek, P. D. Ye, Phosphorene, An unexplored 2D semiconductor with a high hole mobility, ACS Nano. 8 (2014) 4033–4041. https://doi.org/10.1021/nn501226z

[31]   A. C.Gomez, L. Vicarelli, E. Prada, J. O. Island, K. L. N.Acharya, S. Blanter, D. J. Groenendijk, M. Buscema, G. A. Steele, J. V. Alvarez, H. Zandbergen, J. J. Palacios, H. S. J. V. D. Jant Isolation and characterization of few-layer black phosphorus, 2D Mater. 1 (2014) 025001. https://doi.org/10.1088/2053-1583/1/2/025001

[32]   J. Miao, L. Cai, S. Zhang, J. Nah, J. Yeom, C. Wang, Air-stable humidity sensor using few-layer black phosphorus, ACS Appl. Mater. Interfaces. 9 (2017) 10019–10026. https://doi.org/10.1021/acsami.7b01833

[33]   Y. Xu, Y. S. Ang, L. Wu, L. K. Ang, High sensitivity of surface plasmon resonance sensor based on two-dimensional MXene and transition metal

dichalcogenide: A theoretical study, Nanomaterial (Basel). 9 (2019) 165. https://doi.org/10.3390/nano9020165

[34]   L. Wu, Q. You, Y. Shan, S. Gan, Y. Zhao, X. Dai, Y. Xiang, Few-layer $Ti_3C_2T_x$ MXene: A promising surface plasmon resonance biosensing material to enhance the sensitivity, Sens. Actuators B Chem. 277 (2018) 210–215. https://doi.org/10.1016/j.snb.2018.08.154

[35]   K. Hantanasirisakul, M. Q. Zhao, P. Urbankowski, J. Halim, B. Anasori, S. Kota, C. E. Ren, M. W. Barsoum, Y. Gogotsi, Fabrication of $Ti_3C_2T_x$ MXene transparent thin films with tunable optoelectronic properties, Adv. Electron. Mater. 2 (2016) 1600050. https://doi.org/10.1002/aelm.201600050

[36]   S. J. Kim, H. J. Koh, C. E. Ren, O. Kwon, K. Maleski, S. Y. Cho, B. Anasori, C. K. Kim, Y. K. Choi, J. Kim, Y. Gogotsi, H. T. Jung, Metallic $Ti_3C_2T_x$ MXene gas sensors with ultrahigh signal-to-noise ratio, ACS Nano. 12 (2018) 986-993. https://doi.org/10.1021/acsnano.7b07460

[37]   J. Zhu, E. Ha, G. Zhao, Y. Zhou, D. Huang, G. Yue, L. Hu, N. Sun, Y. Wang, L. Y. S. Lee, Recent advance in MXenes: A promising 2D material for catalysis, sensor and chemical adsorption, Coord. Chem. Rev. 352 (2017) 306–327. https://doi.org/10.1016/j.ccr.2017.09.012

[38]   Q. H. Wang, K. K. Zadeh, A. Kis, J. N. Coleman, M. S. Strano, Electronics and optoelectronics of two-dimensional transition metal dichalcogenides, Nature Nanotechnology 7 (2012) 699–712. doi.org/10.1038/nnano.2012.193

[39]   Q. Ouyang, S. Zeng, L. Jiang, L. Hong, G. Xu, X. Q. Dinh, J. Qian, S. He, J. Qu, P. Coquet, K. T. Yong, Sensitivity enhancement of transition metal dichalcogenides/silicon nanostructure-based surface plasmon resonance biosensor, Sci. Rep. 6 (2016) 28190. https://doi.org/10.1038/srep28190

[40]   L. Yan, F. Zhao, S. Li, Z. Hu, Y. Zhao, Low-toxic and safe nanomaterials by surface-chemical design, carbon nanotubes, fullerenes, metallofullerenes, and graphenes. Nanoscale. 3 (2011) 362−382. https://doi.org/10.1039/c0nr00647e

[41]   W. Z. Teo, E. L. K. Chng, Z. Sofer, M. Pumera, Cytotoxicity of exfoliated transition-metal dichalcogenides (MoS_2, WS_2, and WSe_2) is lower than that of graphene and its analogues, Chem. Eur. J. 20 (2014) 9627−9632. https://doi.org/10.1002/chem.201402680

[42] G. Yang, C. Zhu, D. Du, J. Zhu, Y. Lin, Graphene-like two-dimensional layered nanomaterials: applications in biosensors and nanomedicine, Nanoscale. 7 (2015) 14217-14231. https://doi.org/10.1039/C5NR03398E

[43] K. K. Zadeh, J. Z. Ou, T. Daeneke, A. Mitchell, T. Sasaki, M.S. Fuhrer, Two dimensional and layered transition metal oxides, Applied Materials Today. 5 (2016) 73-89. https://doi.org/10.1016/j.apmt.2016.09.012

[44] R. M. Kong, L. Ding, Z. Wang, J. You, F. Qu, A novel aptamer-functionalized MoS_2 nanosheet fluorescent biosensor for sensitive detection of prostate specific antigen, Anal. Bioanal. Chem. 407 (2015) 369–377. https://doi.org/10.1007/s00216-014-8267-9

[45] R. W. Wood, On a remarkable case of uneven distribution of lighting a diffraction grating spectrum, Phil. Magm. 4 (1902) 396 – 402. https://doi.org/10.1080/14786440209462857

[46] E. Kretschmann, H. Raether, Radiative decay of non-radiative surface plasmons excited by light, Z. Naturforsch. 23A (1968) 2135 – 2136. https://doi.org/10.1515/zna-1968-1247

[47] J. Homola, S. S. Yee, G. Gauglitz, Surface plasmon resonance sensors: review, Sensors and Actuators B. 54 (1999) 3–15. https://doi.org/10.1016/j.proeng.2010.11.071

[48] B. H. Ong, X. Yuan, S. C. Tijn, J. Zhang, H. M. Ng , Optimized layer thickness for maximum evanescent field enhancement of a bimetallic layer surface plasmon resonance biosensor, Sens. Actuator B Chem. 114 (2006) 1028–1034. https://doi.org/ 10.1016/j.snb.2005.07.064

[49] H. Raether, Surface plasmons on smooth and rough surfaces and on gratings, Springer-Verlag, Berlin (1988).

[50] S. Pal, Y. K. Prajapati, J. P. Saini, V. Singh, Sensitivity enhancement of metamaterial based surface plasmon resonance biosensor for near infrared, Optica Applicata. 46 (2016) 131–143, 2016. https://doi.org/10.5277/oa160112

[51] Y. K. Prajapati, S. Pal, J. P. Saini, Effect of metamaterial and silicon layers on performance of surface plasmon resonance biosensor in infrared range, Silicon. 10 (2017) 1451–1460. https://doi.org/ 10.1007/s12633-017-9625-y

[52]   S. Pal, Y. K. Prajapati, J. P. Saini, V. Singh, Resolution enhancement of optical surface plasmon resonance sensor using metamaterial, Photonic sensors. 5 (2015) 330-338. https://doi.org/10.1007/s13320-015-0269-5

[53]   J. A. Crump, Progress in typhoid fever epidemiology, Clinical infectious diseases. 68 (2019) S4–S9. https://doi.org/10.1093/cid/ciy846

[54]   Nilima, A. Kamath, K. Shetty, B. Unnikrishnan, S. Kaushik, S. N. Rai Prevalence, patterns, and predictors of diarrhea: a spatial-temporal comprehensive evaluation in India, BMC public health. 18 (2018) 1288–1288. https://doi.org/ 10.1186/s12889-018-6213-z

[55]   T. Shi, A. Denouel, A.K. Tietjen et al., Global and regional burden of hospital admissions for pneumonia in older adults: A systematic review and meta-analysis, The Journal of infectious diseases. (2019) https://doi.org/10.1093/infdis/jiz053

[56]   S. M. Pradhan, A. P. Rao, S. M. Pattanshetty, A. R. Nilima, Knowledge and perception regarding childhood pneumonia among mothers of under-five children in rural areas of Udupi Taluk, Karnataka: A cross-sectional study, Indian Journal of Health Sciences. 9 (2016) 35. https://doi.org/ 10.4103/2349-5006.183690

[57]   S. Palanisamy, S. Ku, S. M. Chen, Dopamine sensor based on a glassy carbon electrode modified with a reduced graphene oxide and palladium nanoparticles composite, Microchim. Acta., 180 (2013) 1037−1042. https://doi.org/ 10.1007/s00604-013-1028-1

[58]   J. Lavanya, N. Gomathi, High-sensitivity ascorbic acid sensor using graphene sheet/graphene nanoribbon hybrid material as an enhanced electrochemical sensing platform, Talanta. 144 (2015) 655−661. https://doi.org/ 10.1016/j.talanta.2015.07.018

[59]   Z. H. . Sheng, X. Q. . Zheng, J. Y. Xu, W. J. Bao, F. B. Wang, X. H. Xia, Electrochemical sensor based on nitrogen doped graphene: Simultaneous determination of ascorbic acid, dopamine and uric Acid, Biosens. Bioelectron. 34 (2012) 125−131. https://doi.org/10.1016/j.bios.2012.01.030

[60]   L. Wu, H. S. Chu, W. S. Koh, E. P. Li, Highly sensitive graphene biosensors based on surface plasmon resonance, Opt. Express. 18 (2010) 14395–14400. https://doi.org/10.1364/OE.18.014395

[61]   R. Verma, B. D. Gupta, R. Jha, Sensitivity enhancement of a surface plasmon resonance based biomolecules sensor using graphene and silicon layers, Sensors and Actuators B. 160 (2011) 623– 631. https://doi.org/10.1016/j.snb.2011.08.039

[62]   S. H. Choi, Y. L. Kim, K. M. Byun, Graphene-on-silver substrates for sensitive surface plasmon resonance imaging biosensors, Opt. Express. 19 (2011) 458–466. https://doi.org/10.1361/OE.19.000458

[63]   P. K. Maharana, R. Jha, Chalcogenide prism and graphene multilayer based surface plasmon resonance affinity biosensor for high performance, Sensors and Actuators B. 169 (2012) 161– 166. https://doi.org/10.1016/j.snb.2012.04.051

[64]   J. A. Kim, T. Hwang, S. R. Dugasani, R. Amin, R. Kulkarni, S. H. Park and T. Kim, Graphene based fiber optic surface plasmon resonance for bio-chemical sensor applications, Sensors and Actuators B. 187 (2013) 426-433. https://doi.org/ 10.1016/j.snb.2013.01.040

[65]   R. Galatus, l. Szolga, E. Voiculescu, Sensitivity enhancement of a D-shape SPR-pof low-cost sensor using graphene, International Journal of Education and Research. 1 (2013).

[66]   P. K. Maharana, R. Jha, S. Palei, Sensitivity enhancement by air mediated graphene multilayer based surface plasmon resonance biosensor for near infrared, Sensors and Actuators B. 190 (2014) 494–501. https://doi.org/10.1016/j.snb.2013.08.089

[67]   A. Verma, A. Prakash, R. Tripathi, Performance analysis of graphene based surface plasmon resonance biosensors for detection of pseudomonas-like bacteria, Optical and Quantum Electronics. 47 (2015) 1197-1205. https://doi.org/10.1007/s11082-014-9976-1

[68]   A. Verma, A. Prakash, R. Tripathi, Sensitivity enhancement of surface plasmon resonance biosensors using graphene and air gap, Optics Communications. 357 (2015) 106- 112. https://doi.org/10.1016/j.optcom.2015.08.076

[69]   S. Pal, A. Verma, Y. K. Prajapati, J. P. Saini., Influence of black phosphorous on performance of surface plasmon resonance biosensor, Optical and Quantum Electronics. 49 (2017) 403. https://doi.org/ 10.1007/s11082-017-1237-7

[70]   J. Pollet, F. Delport, K. P. F. Janssen, K. Jans, G. Maes, H. Pfeiffer, M. Wevers, J. Lammertyn, Fiber optic SPR biosensing of DNA hybridization and DNA-protein

interactions, Biosensors and Bioelectronics. 25 (2009) 864-869.
https://doi.org/10.1016/j.bios.2009.08.045

[71] S. Pal, A. Verma, S. Raikwar, Y. K. Prajapati, J. P. Saini, Detection of DNA hybridization using black phosphorus-graphene coated surface plasmon resonance Sensor, Applied Physics A. 124 (2018) 394. https://doi.org/10.1007/s00339-018-1804-1

[72] B. Meshginqalam, J. Barvestani, Aluminum and phosphorene based ultrasensitive SPR biosensor. Optical Materials. 86 (2018) 119–125. https://doi.org/10.1016/j.optmat.2018.10.003

[73] S. Zenga, S. Hub, J. Xia, T. Anderson, X. Q. Dinh, X. M. Meng, P. Coquet, K. T. Yong, Graphene–MoS$_2$ hybrid nanostructures enhanced surface plasmon resonance biosensors, Sensors and Actuators B. 207 (2015) 801–810. https://doi.org/10.1016/j.snb.2014.10.124

[74] N. A. Jamil, P. S. Menon, F. A. Said, K. A. Tarumaraja, G. S. Mei, B. Y. Majlis, Graphene based surface plasmon resonance urea biosensor using Kretschmann configuration, IEEE Regional Symposium on Micro and Nanoelectronics (RSM) Batu Ferringhi. (2017) 112-115. https://doi.org/ 10.1109/RSM41573.2017

[75] L Wu, J. Guo, Q. Wang, S. Lu, X. Dai, Y. Xiang, D. Fan, Sensitivity enhancement by using few-layer black phosphorus-graphene/TMDCs heterostructure in surface plasmon resonance biochemical sensor, Sensors and Actuators B: Chemical. 249 (2017) 542-548. https://doi.org/ 10.1016/j.snb.2017.04.110

[76] S. Pal, A. Verma, Y. K. Prajapati, J. P. Saini, Sensitivity enhancement using silicon-black phosphorus-TDMC coated surface plasmon resonance biosensor, IET Optoelectronics. 13 (2019) 196-201. https://doi.org/10.1049/iet-opt.2018.5023

[77] B. Meshginqalam, J. Barvestani, Performance Enhancement of SPR biosensor based on phosphorene and transition metal dichalcogenides for sensing DNA hybridization, IEEE Sensors Journal. 18 (2018) 7537-7543. https://doi.org/ 10.1109/JSEN.2018.2861829

[78] A. Srivastava, A. Verma, R. Das, Y. K. Prajapati, A theoretical approach to improve the performance of SPR biosensor using MXene and black phosphorus, Optik. 203 (2019) 163430. https://doi.org/10.1016/j.ijleo.2019.163430

[79]   N. A. S.Omar, Y. W. Fen, S. Saleviter, W. Daniyal, N. Anas, N. S. M. Ramdzan, M. D. A. Roshidi, Development of a graphene-based surface plasmon esonance optical sensor chip for potential biomedical application, Materials (Basel, Switzerland). 12 (2019) 1928. https://doi.org/10.3390/ma12121928

[80]   G. Kaur, A. Paliwal, M. Tomar, V. Gupta, Detection of Neisseria meningitidis using surface plasmon resonance based DNA biosensor, Biosensors and Bioelectronics. 78 (2016) 106–110. https://doi.org/10.1016/j.bios.2015.11.025

[81]   S. Pal, Y. K. Prajapati, J. P. Saini, Influence of graphene's chemical potential on SPR biosensor using ZnO for DNA hybridization, Opt. Rev., 27 (1) (2019) 57-64. https://doi.org/10.1007/s10043-019-00564-w.

[82]   Y. J. Lei, Sensors for toxic gas detection, Platinum Metals Rev. 37 (1993) 146-150.

[83]   A. Paliwal, A. Sharma, M. Tomar, V. Gupta, Carbon monoxide (CO) optical gas sensor based on ZnO thin films, Sensors and Actuators B: Chemical. 250 (2017) 679-685. https://doi.org/10.1016/j.snb.2017.05.064

[84]   E. D. Gaspera, A. Martucci, Sol-Gel thin films for plasmonic gas sensors, Sensors. 15 (2015) 16910-16928. https://doi.org/10.3390/s150716910

[85]   H. D. Wiemhöfer, W. Göpel, Fundamentals and principles of potentiometric gas sensors based upon solid electrolytes, Sensors and Actuators B: Chemical. 4 (1991) 365-372. https://doi.org/10.1016/0925-4005(91)80137-9

[86]   C. Wang, L. Yin, L. Zhang, D. Xiang, R. Gao, Metal oxide gas sensors: Sensitivity and influencing factors, Sensors (Basel). 10 (2010) 2088–2106. https://doi.org/10.3390/s100302088

[87]   W. P. Jakubik, Surface acoustic wave-based gas sensors, Thin Solid Films. 520 (2011) 986-993. https://doi.org/10.1016/j.tsf.2011.04.174

[88]   J. Yanga, L. Zhoua, J. Huanga, C. Taob, X. Lic, W. Chen, Sensitivity enhancing of transition mode long-period fiber grating as methane sensor using high refractive index polycarbonate/cryptophane A overlay deposition, Sensors and Actuators B. 207 (2015) 477–480. https://doi.org/10.1016/j.snb.2014.10.013

[89]   W. Wei, J. Nong, G. Zhang, L. Tang, X. Jiang, N. Chen, S. Luo, G. Lan, Y. Zhu, Graphene-based long-period fiber grating surface plasmon resonance sensor for high-sensitivity gas sensing, Sensors. 17 (2017) 2. https://doi.org/10.3390/s17010002

[90] T. Srivastava, A. Purkayastha, R. Jha, Graphene based surface plasmon resonance gas sensor for terahertz, Opt Quant Electron. 48 (2016) 334. https://doi.org/10.1007/s11082-016-0462-9

[91] A. Purkayastha, T. Srivastava, R. Jha, Ultrasensitive THz-plasmonics gaseous sensor using doped graphene, Sens. Actuators B Chem. 227 (2016) 291–295. https://doi.org/10.1016/j.snb.2015.12.055

[92] P. K. Maharanaa, R. Jha, P. Padhy, On the electric field enhancement and performance of SPR gas sensor based on graphene for visible and near infrared, Sensors and Actuators B. 207 (2015), 117-122. https://doi.org/10.1016/j.snb.2014.10.006

[93] J. B. Maurya, S. Raikawar, Y. K. Prajapati, J. P. Saini, A Silicon-black phosphorous based surface plasmon resonance sensor for the detection of NO_2 gas, Optik. 160 (2018) 428-433. https://doi.org/10.1016/j.ijleo.2018.02.002

[94] L. Wu, Q. Wang, B. Ruan, J. Zhu, Q. You, X. Dai, Y. Xiang, High performance lossy-mode resonance sensor based on few-layer black phosphorus, The Journal of Physical Chemistry C. 122 (2018) 7368-7373. https://doi.org/ 10.1021/acs.jpcc.7b12549

[95] Y. Singh S. K. Raghuwanshi, Sensitivity enhancement of the surface plasmon resonance gas sensor with black phosphorus, IEEE sensor letters. 3 (2019) 1-4. https://doi.org/10.1109/LSENS.2019.2954052

[96] H. Wang, H. Zhang, J. Dong, S. Hu, W. Zhu, W. Qiu, H. Lu, J. Yu, H. Guan, S. Gao, Z. Li, W. Liu, M. He, J. Zhang, Z. Chen, Y. Luo, Sensitivity-enhanced surface plasmon resonance sensor utilizing a tungsten disulfide (WS_2) nanosheets overlayer, Photonics Research. 6 (2018) 485-491. https://doi.org/ 10.1364/PRJ.6.000485

[97] G. A. Asres, J. J. Baldoví, A. Dombovari, T. Järvinen, G. S. Lorite, M. Mohl, A. Shchukarev, A. P. Paz, L. Xian, J. P. Mikkola, A. L. Spetz, H. Jantunen, Á. Rubio, K. Kordas, Ultrasensitive H_2S gas sensors based on p-type WS_2 hybrid materials, Nano Research. 11 (2018) 4215–4224. https://doi.org/10.1007/s12274-018-2009-9

[98] A. Sharma, A. Pandey, Blue Phosphorene/MoS_2 heterostructure based SPR sensor with enhanced Sensitivity, IEEE Photonics Technology Letters. PP (99) 1-1 (2018) https://doi.org/10.1109/LPT.2018.2803747

[99] E. Lee, A. V. Mohammadi, B. C. Prorok, Y. S. Yoon, M. Beidaghi, D. J. Kim, Room temperature gas-sensing of two-dimensional titanium carbide (MXene), ACS

Applied Materials & Interfaces. 9 (2017) 37184-37190. https://doi.org/
10.1021/acsami.7b11055

[100] L. Lorencová, T. Bertok, E. Dosekova, A. Holazová, D. Paprckova, A.
Vikartovská, V. Sasinková, J. Filip, P. Kasák, M. Jerigová, Electrochemical
performance of $Ti_3C_2T_x$ MXene in aqueous media: towards ultrasensitive H_2O_2
sensing, Electrochim. Acta. 235 (2017) 471–479. https://doi.org/
10.1016/j.electacta.2017.03.073

[101] S. K. Mishra, D. Kumari, B. D. Gupta, Surface plasmon resonance based fiber
optic ammonia gas sensor using ITO and polyaniline, Sensors and Actuators, B:
Chemical. 171–172 (2012) 976–983. https://doi.org/10.1016/j.snb.2012.06.013

[102] R. Tabassum, S. K. Mishra, B. D. Gupta, Surface plasmon resonance-based fiber
optic hydrogen sulphide gas sensor utilizing Cu–ZnO thin films, Phys. Chem. Chem.
Phys. 15 (2013) 11 868–11 874. https://doi.org/10.1039/C3CP51525G

[103] A. Paliwal, A. Sharma, M. Tomar, V. Gupta, Carbon monoxide (CO) optical gas
sensor based on ZnO thin films. Sensors and Actuators, B: Chemical. 250 (2017)
679–685. https://doi.org/ 10.1016/j.snb.2017.05.064

[104] T. S. Kim, Y. B. Kim, K. S. Yoo, G. S. Sung, H. J. Jung, Sensing characteristics of
dc reactive sputtered WO_3 thin films as an NO_x gas sensor, Sensors and Actuators B.
62 (2000) 102-108. https://doi.org/10.1016/S0925-4005(99)00360-3

[105] E. LeBlanc, L. P. Camby, G. Thomas, R. Gibert, M. Primet, P. Gelin, NO_x
adsorption onto dehydroxylated or hydroxylated tin dioxide surface. Application to
SnO_2-based sensors, Sensors and Actuators B. 62 (2000) 67. https://doi.org/
10.1016/S0925-4005(99)00376-7

[106] C. Cantalini, W. Wlodarsky, H. T. Sun, M. Z. Atashbar, M. Passacantando, S.
Santucci, NO_2 response of In_2O_3 thin film gas sensors prepared by sol–gel and
vacuum thermal evaporation techniques, Sensors and Actuators B. 65 (2000) 101-104.
https://doi.org/10.1016/S0925-4005(99)00439-6.

[107] D. D. Lee, D. S. Lee, Environmental gas sensors, IEEE Sensors Journal. 1 (2001)
214. https://doi.org/10.1109/JSEN.2001.954834

[108] G. Harsanyi, Polymer films in sensor applications: a review of present uses and
future possibilities, Sensor Review. 20 (2000) 98-105. https://doi.org/
10.1108/02602280010319169

[109] D. Li, Y. Jiang, Z. Wu, X. Chen, Y. Li, Self-assembly of polyaniline ultrathin films based on doping-induced deposition effect and applications for chemical sensors, Sensors and Actuators B. 66 (2000) 125-127. https://doi.org/ 10.1016/S0925-4005(00)00315-4.

[110] T. A. Chen, X. Wu, R. D. Rieke, Regiocontrolled synthesis of Poly(3-alkylthiophenes) mediated by rieke Zinc: Their characterization and solid-state properties, Journal of American Chemical Society. 117 (1995) 233-244. https://doi.org/ 10.1021/ja00106a027

[111] J. C. Solı´s, E. D. L. Rosa, E. P. Cabrera, Absorption and refractive index changes of poly (3-octylthiophene) under NO_2 gas exposure, Optical Materials. 29 (2006) 167–172. https://doi.org/10.1016/j.optmat.2005.07.009

Toxic Gas Sensors and Biosensors
Materials Research Foundations **92** (2021) 107-138

Materials Research Forum LLC
https://doi.org/10.21741/9781644901175-4

Chapter 4

MXenes for Gas and Biological Sensor

Sze-Mun Lam[1,2,3,4]*, Zeeshan Haider Jaffari[4], Yit-Thai Ong[5], Jin-Chung Sin[1,2,3,5], Hua Lin[1,2,3], Haixiang Li[1,2,3], Honghu Zeng[1,2,3]*

[1]College of Environmental Science and Engineering, Guilin University of Technology, Guilin 541004, China

[2]Guangxi Key Laboratory of Theory and Technology for Environmental Pollution Control, Guilin University of Technology, Guilin 541004, China

[3]Collaborative Innovation Center for Water Pollution Control and Water Safety in Karst Area, Guilin University of Technology, Guilin 541004, China

[4]Department of Environmental Engineering, Faculty of Engineering and Green Technology, Universiti Tunku Abdul Rahman, 31900 Kampar, Perak, Malaysia

[5]Department of Petrochemical Engineering, Faculty of Engineering and Green Technology, Universiti Tunku Abdul Rahman, 31900 Kampar, Perak, Malaysia

lamsm@utar.edu.my; zenghonghu@glut.edu.cn *

Abstract

Recent advancement of two dimensional MXene nanomaterial offers promise in gases and biosensor areas owing to its large surface area, high thermal conductivity, remarkable safety and excellent catalytic activity traits. The current chapter aimed to review the fundamental and technological aspects of MXenes, including myriad synthesis techniques and structural as well as electronic characteristics of these compounds. The features elucidated in the subsequent sections, examined by both theoretical and experimental approaches and potentialities of MXenes in the gas removal and biosensor applications. Several challenges and exciting future opportunities of this research platform are lastly summarized.

Keywords

MXene, Two-dimensional, Gas, Environmental, Biosensor

Contents

1. Introduction

Since the graphene discovery, the family of two-dimensional (2D) nanomaterials is one of the most investigated materials [1,2]. The 2D nanomaterials are commonly applied in environmental remediation, energy storage, superconductors, capacitors and sensors because of their unique physical and chemical characteristics. Typically, one dimension of 2D nanomaterials is only restrained to the single/few layers of atoms (< 5 nm), while the second dimension might be higher than 100 nm [3–8]. Until now, 18 different kind of 2D materials beyond graphene were invented: transition metal dichalcogenides (TMDs) [9–11], hexagonal boron nitride (h-BN) [12–14], black phosphorus (BP) [15,16], graphitic carbon nitride (g-C_3N_4) [17], metal phosphorus trichalcogenides [18], III–VI layered semiconductors [19], double hydroxides (LDHs) [20,21], covalent–organic frameworks (COFs) [22] and polymers [23–25], metals [26], perovskites [27,28] and niobates [29], metal halides [30], some metal–organic frameworks (MOFs) [2,31,32], layered metal oxides, transition metal oxyhalides[33], non-layer structured metal oxides

[34], non-layer structured metal chalcogenides[35], silicates and hydroxides (clays) [36] and transition metal carbides and/or nitrides (MXenes) [37,38].

Among the different classes of 2D nanomaterials, the MXene materials are extremely young and rapidly growing. Until now, over 70 different types of MXenes were successfully synthesized and reported in the literatures [37–40]. The MXene materials are successfully synthesized through the selective etching of "A" in the "MAX" phase. Normally, $M_{n+1}AX_n$ is the formula of the MAX phases, where M is the transition metals, A is the *sp*-elements and X is either carbon (C) and/or nitrogen (N) [41]. Fig. 1 [38] presents the structures of M_2X, M_2X_2 and M_3X_2 MXenes. Ti_3C_2 was the first-ever reported MXene material, which was synthesized at room temperature through the selective etching of Ti_3AlC_2 layered ternary carbide with concentrated hydrofluoric acid (HF) [42]. In most of the MXene materials, the "M" element contains more than one atom and two different structural forms: ordered phases and solid solutions. The arbitrary alignment of two transition metals were witnessed in the M layers solid solution structural form, while in the ordered phase structure single/double layers of first transition metal were sandwiched among the layers of second transition metal in 2D structure.

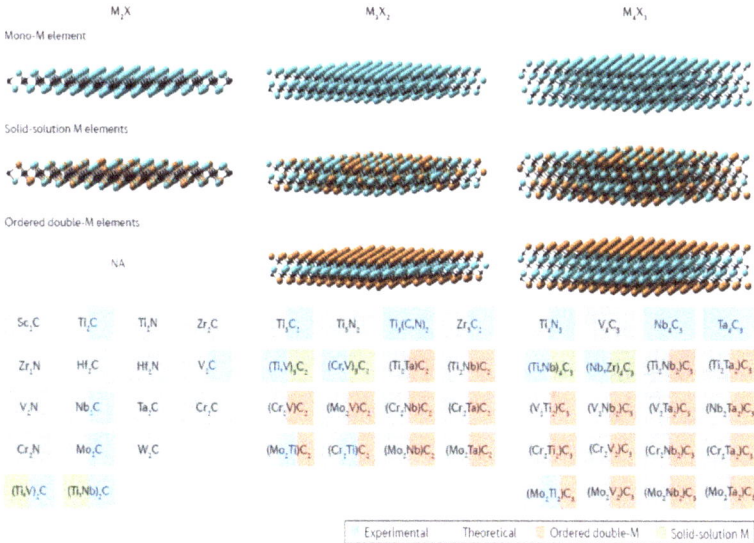

Fig. 1 MXene materials reported so far with three formulas: M_2X, M_3X_2 and M_4X_3, where M is the transition metals and X is either C and/or N [38].

In this book chapter, the removal of gasses and biosensing applications of MXene materials are systematically reviewed. The synthesis methods for multilayers and single/few-layered MXenes are first discussed and followed by their surface functionalization and electronic properties. In the next section, their practical applications towards the removal of gasses through various techniques, such as membrane filtration, adsorption and photocatalysis are explained in detail. The sensing and biosensing applications towards different gasses, macromolecules and cell detection are also explicated. Finally, future prospects together with challenges towards the removal of gases and biosensing applications of MXenes are highlighted.

2.   Preparation of MXenes

Since the discovery of $Ti_3C_2T_x$ [42], the synthesis techniques for MXene materials have garnered tremendous attention. Various novel techniques have been established to fabricate MXene materials.

2.1   Multi-layered stacked MXenes

Naguib et al. [42] synthesized the MXene material for the first time by dipping Ti_3AlC_2 powder in an HF solution with the concentration of 50 wt% for 2 hr. Afterwards, they washed the suspension several times with distilled water, and the obtained powder was separated using a centrifugation process. The proposed mechanism suggested that the MXene synthesis followed a selective etching of Al atomic layers. Such discretionary etching was only conceivable due to the easier cleavage of Ti-Al bond compared with Ti-C bond as presented in Fig. 2(a) [42]. The superior Ti-Al bond etching in the Ti_3AlC_2 phase was summarized by the following chemical reactions.

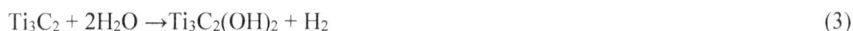

$$Ti_3AlC_2 + 3HF \rightarrow AlF_3 + 1.5\ H_2 + Ti_3C_2 \tag{1}$$

$$Ti_3C_2 + 2HF \rightarrow Ti_3C_2F_2 + H_2 \tag{2}$$

$$Ti_3C_2 + 2H_2O \rightarrow Ti_3C_2(OH)_2 + H_2 \tag{3}$$

The Eqs. (2) and (3) suggested the formation of –F and –OH surface groups on the MXene, which ultimately ended up as the surface termination of exfoliated Ti_3C layers. It must be acclaimed that both Eqs. (2) and (3) were abridged based on the assumption that the –F and –OH was the terminations respectively, yet actually they were possibly a consolidation of both as presented in Figs. 2 [42] (b) and (c). Various novel MXene materials, for example, V_2CT_x [43], $Ta_4C_3T_x$ [44], Mo_2CT_x [45,46] were successfully fabricated using the selective HF etching technique. Many solid solutions, for instance,

$(Ti_{0.5}Nb_{0.5})_2CT_x$ [44], $(Nb_{0.8}Ti_{0.2})_4C_3T_x$ [47] and $(V_{0.5}Cr_{0.5})_3C_2T_x$ [44] reported that the numerous elements were randomly dispersed on the M-site. Anasor et al. [48] were the first who revealed the fabrication of double-M-element $Mo_2Ti_2C_3T_x$ MXenes, which further expanded the choices of chemistry, structure and ultimately beneficial applications. Until now, Hf-based and Zr-based MXenes were not successfully synthesized using Hf-based and Zr-based MAX phase etching protocol, which hampered the addition to the MXene family. Zhou et al. [49] fabricated the layered $Zr_3C_2T_x$ using extended HF etching to $Zr_3Al_3C_5$. In another study, Zhou et al. [50] also fabricated $Hf_3C_2T_x$ through selective etching Al_4C_4 from $Hf_3Al_4C_6$.

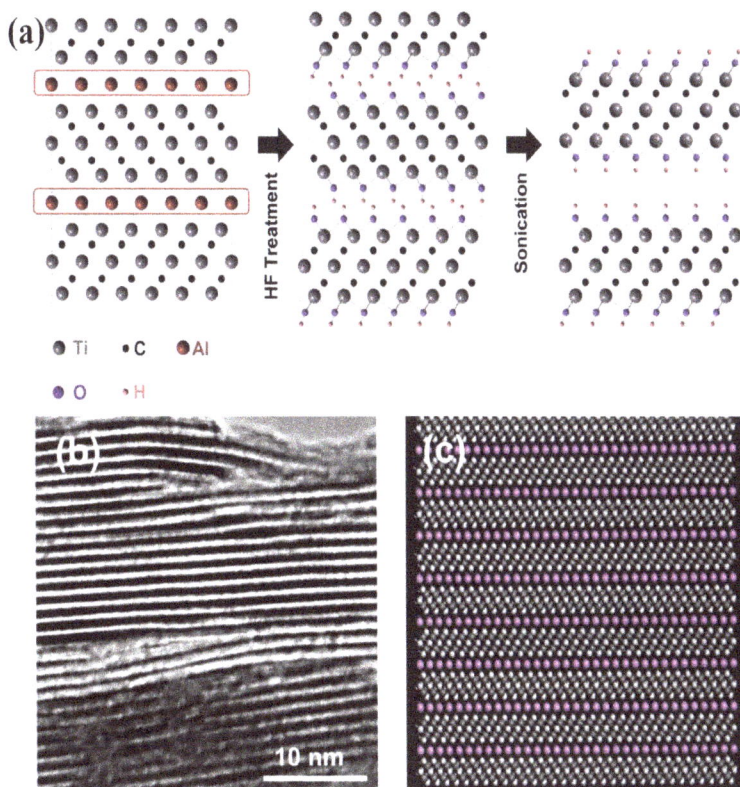

Fig. 2 (a) Schematic diagram of the Ti_3AlC_2 MXene exfoliation, (b) TEM image of stacked TiC_2T_x MXene layers, and (c) Simulated Li-intercalated model structure of multilayers Ti_3C_2 ($Ti_3C_2Li_2$) [42].

The application of concentrated HF in the production of MXene materials restricted their practical applications due to its harmful and non-ecofriendly effects. Hence, it was highly important to develop safe and fast alternative synthesis routes. Ghidiu et al. [51] fabricated Ti_3AlC_2 material by utilizing relatively less harmful lithium fluoride (LiF) salt and hydrochloric acid (HCl). They obtained high yield MXene product using a novel fast and facile method, which performed etching and intercalation at the same time. Shahzad et al. [52] altered the synthesis method by applying a higher molar ration of 7.5:1 (instead of 5:1) for $LiF:Ti_3AlC_2$ and sonication step was eliminated, which allowed extra intercalation of Li^+ ions. The etching solution in the current process was produced by mixing LiF (1 g) in 6 M HCl solution. Afterwards, 1 g of Ti_3AlC_2 was added in the etching solution and the reaction was performed for 24 h at 35 °C. Lipatov et al. [53] further investigated the quality of the synthesized flakes by changing the LiF to Ti_3AlC_2 ratios. Their results presented that with the molar ratio of 5:1 (LiF: Ti_3AlC_2) the exfoliated flake diameters were ranged between 200−500 nm, while the diameters were recorded between 4−15 nm with a molar ratio of 7.5:1.

In another work by Wang et al. [54], Ti_3AlC_2 was immersed in NH_4F solution to synthesize $Ti_3C_2T_x$ in a Teflon-lined stainless steel autoclave for 24 h at 150 °C. This synthesis method was more complex as compared to others. However, they eluded the HF utilization, which made the product less harmful to the environment. The complexities in the synthesis method was reduced by Feng et al. [55]. They introduced a simple and less hazardous process using milder etchants, such as $NaHF_2$, NH_4HF_2 and KHF_2, which instantaneously intercalate cations (Na^+, NH_4^+, K^+) and successfully etched the Al layers. The synthesis approach was explained using the following chemical reactions:

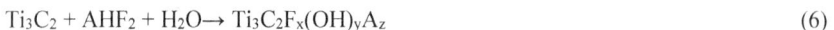

$$Ti_3AlC_2 + AHF_2 \rightarrow A_xAlF_y + AlF_3 + Ti_3C_2 + H_2 \tag{4}$$

$$AlF_3 + aH_2O \rightarrow AlF_3.aH_2O \tag{5}$$

$$Ti_3C_2 + AHF_2 + H_2O \rightarrow Ti_3C_2F_x(OH)_yA_z \tag{6}$$

Until now, most of the synthesis works of nitride-based MXenes using HF etching were ineffective. The estimated cohesive energies for $Ti_{n+1}N_n$ were lower as compared with the $Ti_{n+1}C_n$, while the synthesizing energies of $Ti_{n+1}N_n$ through $Ti_{n+1}AlN_n$ were reported to be higher as compared with $Ti_{n+1}C_n$ through $Ti_{n+1}AlC_n$. The least cohesion energy suggested that the synthesized MXene had lower structural stability, while the higher MXene synthesizing energy suggested the corresponding Al-containing MAX phase needed extra energy for extraction. In other words, the Al elements were bonded more tightly in

Toxic Gas Sensors and Biosensors   Materials Research Forum LLC
Materials Research Foundations **92** (2021) 107-138   https://doi.org/10.21741/9781644901175-4

$Ti_{n+1}AlN_n$ than those of the $Ti_{n+1}AlC_n$, which might be another reason for not been synthesized the nitride-based MXenes using HF etching [37,38]. Urbankowski et al. [56] synthesized $Ti_4N_3T_x$ using a molten salt method by mixing equal masses of Ti_4AlN_3 powder and different salts of fluoride (KF, NaF, LiF) for 30 min at 550 °C under Ar atmosphere as shown in Fig. 3 [56].

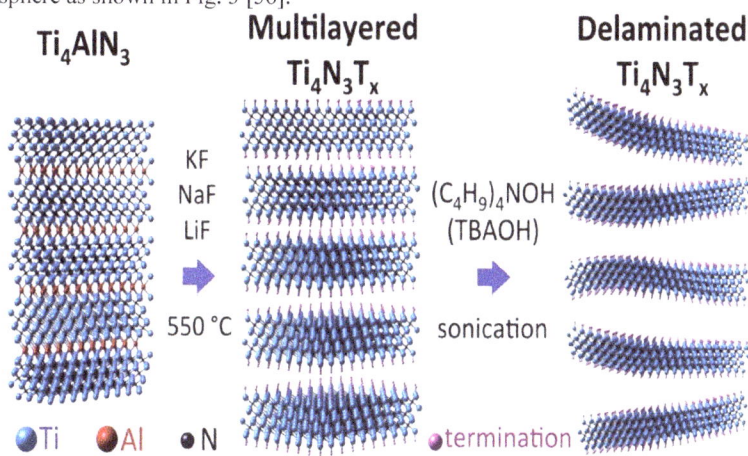

Fig. 3 Schematic diagram of the $Ti_4N_3T_x$ synthesis using a molten salt technique under Ar atmosphere at 550°C and followed by multilayered MXene delamination through TBAOH [56].

All the MXene materials as aforementioned were synthesized by etching of Al and Si elements from MA phases with the surface groups of –F, –OH and –O. Nonetheless, the pure $M_{n+1}X_n$ and $M_{n+1}X_nT_x$ with surface groups had different properties. Hence, it was necessary to produce some novel techniques to synthesize pure MXene materials. Xu et al. [57] proposed a new chemical vapor deposition method to synthesize extremely pure - Mo_2C MXene on Mo/Cu foil at above 1085 °C under CH_4 atmosphere. The extremely higher synthesis temperature can melt the Cu and formed Mo-Cu alloys. The Xu and his team proposed that the Mo atoms were diffused through the interface in the Cu liquid and produced Mo_2C crystals by reacting with the CH_4 gas.

2.2   Single/Multi-layered MXenes

The replacement of firm M–A bonds with the weak forces, like *Van der Waals* forces and hydrogen bonds and breakage of multi-layered MXene materials into the few/single layer

Toxic Gas Sensors and Biosensors   Materials Research Forum LLC
Materials Research Foundations **92** (2021) 107-138   https://doi.org/10.21741/9781644901175-4

were also done successfully. Lukatskaya et al. [58] presented the impulsive various cations intercalation, such as Mg^{2+}, Na^+, NH_4^+, Al^{3+} and K^+ from the respective salt solutions among Ti_3C_2MXene layers. The intercalation would enhance the MXenes layer spacing, which permitted the large metal ions and particles to intercalate or deintercalate. Initially, it was difficult to control the interlayer spacings, which restricted the selective intercalation for numerous metallic ions. Nevertheless, Luo et al. [59] synthesized controllable layer spacings Ti_3C_2MXene for the first time. Their investigation reports found that various lengths of hydrophobic alkyl chains were initially intercalated with the cationic surfactants. The Ti_3C_2 was then inserted into the cationic solution and the surfactants was instantaneously intercalate among the negatively charged Ti_3C_2 layers due to the electrostatic interaction and led to an increase among the Ti_3C_2 layer spacings.

One should also take note that the weak forces among the layers of $M_{n+1}X_n$ allowed the sonication to increase interlayer spacings among the few layers. Nevertheless, low delaminated yields restricted the practical applications, where high delamination among layers were necessary. Mashtalir et al. [60] presented the immense delamination in the $Ti_3C_2T_x$ layers through dimethyl sulfoxide (DMSO) intercalation at 25 °C for 24 h of vigorous magnetic stirring and followed by 6 h of sonication in a water bath. This circumstance might happen due to the H_2O capillary condensation from the air in the interlayer space. It was noteworthy that this technique was only applicable for the $Ti_3C_2T_x$ synthesis as the DMSO was not capable to intercalate during the synthesis of other MXene materials. In another study, Ti_3AlC_2 was treated with the HCl and LiF instead of HF. The report indicated they were again able to synthesize delaminated $Ti_3C_2T_x$ [61].

Several works focused on synthesis of other single or few layered MXenes have been also revealed. Naguib et al. [62] applied large organic bases, such as choline hydroxide and tetrabutylammonium (TBAOH) to synthesize high yield delaminated V_2CT_x, Nb_2CT_x and Ti_3CNT_xMXenes. The as-synthesized MXenes presented higher d-spacing and lower contents of F, which was because of the –O and –OH replacements. Urbankowski et al. [56] also applied TBAOH to synthesized delaminated $Ti_4N_3T_x$ under strong sonication conditions.

Recently, it was also a hot topic to synthesize 2D MXene materials with different morphologies beyond the layer structure to obtained good mechanical and electronic properties. For the first time, Vaughn et al. [63] used selective ring sized intercalating p-phosphonic calix[n]arenes (PCXn, where n was equal to the 4, 5, 6 and 8) to fabricate the Ti_2CT_x materials with plates, sphere, scrolls and crumpled sheets morphologies, respectively as shown in Fig 4 [63]. The microsized sheet and crumpled-up like structures were obtained when the sonication was performed in the presence of PCX4 and PCX5, respectively. The sonication in the presence of PCX6 produced unexpected spherical shaped materials. The biggest phosphonic acid PCX8 assisted the generation of scrolls-like morphologies after the sonication process, which might be due to their exceptional flexibilities and dexterities.

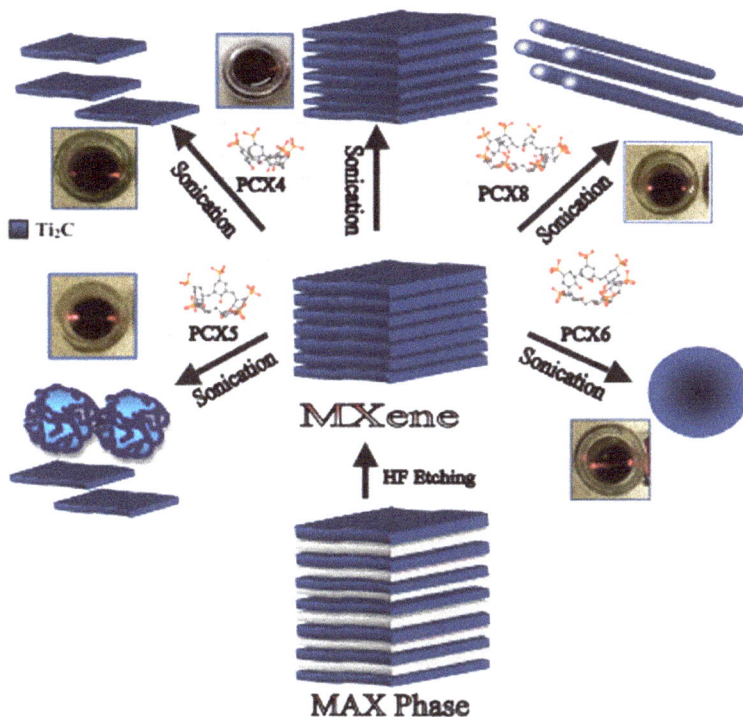

Fig. 4 Schematic diagram summarizing the Ti$_2$C fabrication with different morphologies. The MXenes were fabricated by HF etching, followed by ultrasonication in the presence of PCXn [63].

3.   Surface functionalization and electronic properties of MXenes

The outer surface of the exfoliated MXene materials were usually functionalized with various functional groups (−F, −OH or −O) in an aqueous HF solution environment [39,64–66]. The surface functionalization left substantial effect on the MXenes ion transfer and electronic properties, which were mainly connected with the heterogeneous electron transfer and conductivity processes happening at the surface of MXenes. Hence,

Toxic Gas Sensors and Biosensors   Materials Research Forum LLC
Materials Research Foundations **92** (2021) 107-138   https://doi.org/10.21741/9781644901175-4

the electronic activities of MXene materials were highly dependent on their surface functionalization [39]. For example, all the pristine M_2X-MXenes had the Fermi energy level at transition-metal d orbital, which made them metallic in nature. Some MXene materials also presented the semiconductor properties on functionalization such as Ti_3C_2 had a metallic nature, while $Ti_3C_2OH_2$ and $Ti_3C_2F_2$ were the semiconductors [37,39]. Most of the MXene materials had smaller bandgap energy values, as the p orbitals of X (N/C) were situated underneath the transition metals d orbitals. New energy bands were created lower than the Fermi level on functionalization as a result of hybridization among the d orbital of transition metals and p orbitals of terminating groups. Moreover, the transition metals can easily donate the electrons to the terminating groups and became positively charged due to the smaller electronegativity. This produced the base of various MXene semiconductors, such as Sc_2COH_2 Sc_2CO_2, Sc_2CF_2, Zr_2CO_2, Hf_2CO_2 and Ti_2CO_2, which was only because of the Fermi levels shifting at the center among the transition metals d orbitals and C/N (X) p orbitals. It was also worth mentioning that the metallic to semiconductor nature was dependent on the transition among centered honeycomb (T) and honeycomb (H) phases. The difference in atomic stacking of transition metals was also noted as exhibited in Fig. 5 [67].

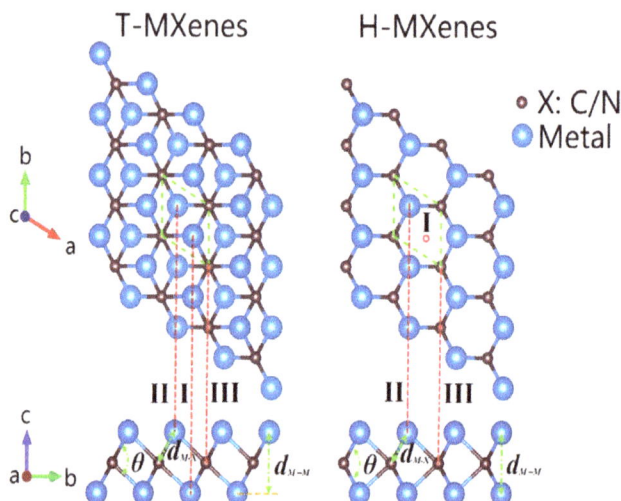

Fig. 5 Atomic structures of T-MXene and H-MXene [67].

4.   MXenes for gas

As mentioned above, MXenes are attractive to versatile applications ranging from catalysis, energy storage to sensors because of their unique and appealing traits. These properties made MXenes serve as promising candidates in environmental remediation, especially gaseous removal. Progress has been made in this application not only through both theoretical and experimental researches, but also accelerated the construction of MXene-based nanoscience in environmental-related technological era.

4.1   Application of MXenes as membranes for gas separation

The emergence of the two-dimensional (2D) MXenes particularly titanium carbide ($Ti_3C_2T_x$) constitutes a breakthrough in the development of membrane for gas separation. Owing to their down to atomic scale thickness, the integration of MXenes as building blocks enables the design of ultrathin layer membrane which could minimize the transport resistance for permeation, while maintaining high selectivity [68,69].

The integration of MXenes in design of membrane can be simply accomplished by blending the exfoliated MXene nanosheet into a continuous matrix to form a so-called mixed matrix membrane [70-72]. Nonetheless, substantial efforts have been devoted to assemble the 2D MXene nanosheet into a laminar structure membrane via a vacuum-assisted filtration method. The laminar structure of MXene layer which composed of stacked MXene nanosheet exploited the advantage of the 2D structure of MXene to form an ultrathin membrane with adequate mechanical properties [73-75]. Meanwhile, the presence of functional group on MXene nanosheets created an interlayer distance in between the lamellar stacked MXene nanosheet. The interlayer spacing as well as the interstitial spacing between the adjacent nanosheets form tortuous transport nanochannels for selective molecule permeation [76]. The ordered lamellar stacked MXene membrane offered a fascinating ideal to fabricate an emerging class of molecular sieving membranes.

The MXene membrane has demonstrated a great prospect in water purification with its remarkably water permeance (>1000 L/(m^2·h·bar)) and high rejection rate to exclude molecule with size down to 2.5 nm [77]. The capability of MXene membrane to screen out much smaller size molecules opens a door for its application for gas separation. The molecular sieving properties of MXene membrane with interlayer spacing of approximately 0.35 nm permits the high permeability of small gas molecules such as hydrogen (H_2) which has a kinetic diameter of 0.29 nm and render a higher selectivity for the permeation of H_2 in separation of H_2/carbon dioxide (CO_2) gas mixtures [73,78]. In addition, the lamellar stacked MXene membrane also demonstrated high temperature tolerance, which can be attributed to the unique covalent, ionic and metallic bonds in

MXene that endowed it to be thermodynamically stable structure at high temperature condition. Such robust structure allowed the application of the MXene membrane in gas mixture separation using at high temperature environment as high as 320 °C [78]. Interestingly, the 2D structure MXenes also possessed tunable physicochemical properties, which allowed the MXene membrane to be tailored for specific separation application. As stated earlier, the synthesis of MXene nanosheets are usually terminated with functional group, which included –O, –H, and –F group. The presence of these functional groups enabled chemical functionalization to tune the interlayer spacing between MXene nanosheet to enhance the selective permeation of specific gas molecules. Shen et al. [79] manipulated the stacking behaviour and interlayer spacing of MXene membrane by functionalizing borate and polyethyelenimine (PEI) molecules within the MXene nanosheets. The borate and PEI molecules firmly interlocked the nanosheet by crosslinked the oxygen moieties and forming hydrogen bonding framework with the MXene nanosheet in order to fine tune the stacking behavior and interlayer spacing for permeation of specific molecules. Both the borate and PEI molecules were found to enhance the affinity of the MXene membrane toward CO_2 permeance and permitted high selectivity for CO_2 permeation from CO_2/ methane (CH_4) gas mixture.

The unique molecular sieving properties of the lamellar stacked MXene membrane offered a promising potential to break the Robinson upper bound or the limitation in permeability-selectivity trade-off, which was the inherent limitation found in conventional polymeric membranes [76,80]. The superior separation capability of MXene membranes in gas separation, which may transcend the state-of-the-art membranes opened up the potential of the MXene membrane to be applied for H_2 purification in methanol reforming reaction as well as CO_2 capture for clean and sustainable power production.

4.2   Application of MXenes in adsorption of gases

In the past few decades, innovation in developing new adsorbent for gas have been mainly focused on 2D materials owing to their unique physical and chemical characteristics. The MXenes being 2D materials inevitably become one of the extensively studied adsorbent materials for gas. The flat sheet structure of MXenes offered high specific surface area, which was essential to supply more available sites for adsorption. In addition, its hydrophilic nature attributed to the abundance of highly active oxygen containing functional group (-OH and -O) generated on the surface after chemical etching process further strengthen its sorption capability for molecular and ionic species [81]. Among the -OH and -O functional group, studies showed that the -OH terminated MXene was found to exhibit stronger gas adsorption compared to -O termintated MXene [82,83].

Both theoretical and experimental studies clearly demonstrated the promising adsorption behavior of MXenes toward CO_2, H_2, CH_4, NH_3 and volatile organic compounds (VOCs). The adsorption behavior of CO_2 and CH_4 mainly depend on physisorption on oxygen terminated MXenes (Ti_2C and Ti_3C_2) [84-86]. For H_2, finding based on first principles simulation revealed that the interaction with MXenes mainly via physisorption of H_2 molecules, Kubas-type interaction between the H_2 molecules with the transition metal and chemisorption of dissociated H atoms [87]. The MXenes appear as promising hydrogen storage medium considering that the maximum hydrogen adsorption capacity of 8.6, 9.0, 7.6 and 8.555 wt.% were estimated to be achieved by Ti_2C [87], Sc_2C [88], Cr_2C [89] and Ti_2N [90], respectively. Apart from that, simulation studies also deduced a stronger interaction in the adsorption of NH_3 on MXenes. The NH_3 was found to be chemisorbed on the O-terminated MXenes, which included Ti_2CO_2, Sc_2CO_2, Zr_2CO_2 and Hf_2CO_2 with an apparent charge transfer [83,91]. Interestingly, the adsorption behaviour of NH_3 can be altered from chemisorption to physisorption by simply injecting electrons into the MXenes [92]. The tunable adsorption behaviour indicates a potential application of MXene as NH_3 sensor and capturer. Besides, MXenes such as $Ti_3C_2T_x$ also demonstrated high selectivity toward adsorption of VOCs, such as acetone, ethanol and propanal. The adsorption of the VOC led to an increase resistance and hinder the charge carrier transport in $Ti_3C_2T_x$ [82,92]. The adsorption behavior rendered the $Ti_3C_2T_x$ superior sensing property for VOC at a very low limit of detection (50-100 ppb) at room temperature [82].

It has been observed that the gas adsorption properties of MXenes can be influenced by selection of etchant during the synthesis of MXenes. Due to corrosive concern, mixture of fluoride salt in HCl solution becomes an alternate etchant to replace HF. The selection of different fluoride salts, such as LiF, NaF, KaF, and NH_4F as an etchant that could influence the ions to absorb on the surface of MXenes and altered their surface characteristic for gas adsorption [86]. Besides that, intercalation also played a critical role in the adsorption properties of MXene. The use of intercalating agents, such as ammonia monohydrate ($NH_3 \cdot H_2O$), dimethyl sulphoxide and urea increased the spacing within the stacked MXene and further exfoliated the MXene to provide larger specific surface area and higher pore volume for gas adsorption [84-86]. As a result, the CH_4 adsorption capacity of Ti_2C intercalated with $NH_3 \cdot H_2O$ was reported to be improved by 45% as compared to Ti_2C without intercalation [85].

4.3 Application of MXenes in photodegradation of gases

As 2D materials, MXenes have promising physicochemical traits in photocatalytic process owing to their large surface area, ease of functionalization and abundant active sites [93,94]. MXenes were however still having several limitations, like poor light absorption and rapid recombination rate of electron-hole pair, which could be ameliorated by tailoring good contact heterojunction interfaces between MXene and other nanoparticles. Novel MXene nano-architectures with good light absorption capability, such as $Ti_3C_2T_x$-TiO_2 can fulfil the quantum efficiency and improve electron-hole pair separation, attaining outstanding photo-redox efficiency in photocatalytic application. Other MXene-oxide systems (eg., MoO_3-MXenes, ZnO-MXenes, Nb_2O_5-MXenes, and etc.) should also be discovered. Those oxides can either be decorated on MXenes can be formed *in situ* through MXenes partial oxidation.

For volatile organic carbons (VOCs) photodegradation, Huang et al. [95] introduced the Ti_3C_2MXene nanoparticles on Bi_2WO_6 nanoplates via an electrostatic attraction route. The Bi_2WO_6-Ti_3C_2MXene showed the maximum $v(CO_2)$ of 72.8 and 85.3 $\mu molg^{-1}h^{-1}$ for the HCHO degradation and CH_3COCH_3. Density functional theory (DFT) calculations indicated that Ti_3C_2 MXene can chemically adsorb HCHO and CH_3COCH_3 molecules, while Bi_2WO_6 only demonstrated weak physical adsorption. The excellent adsorption capability of Ti_3C_2MXene lied on the substantial charge carrier transport between Ti_3C_2MXene and VOCs molecules as witnessed by differential charge densities and Bader charge tests. The resulting Bi_2WO_6-Ti_3C_2MXene achieved effective charge carrier separation in the presence of light irradiation because of the outstanding photogenerated electron-hole pairs separation and favourable VOCs adsorption of Ti_3C_2MXene.

5.   MXene as sensors

In spite of the very small journey, MXene materials had shown to be extremely sensitive and selective materials for sensing applications due to their salient properties. In the present section, various sensing and biosensing-related applications of MXenes were discussed and listed in Table 1 [96-110].

Table 1   MXene-based sensors and their parameters [96-110]

Analyte	MXene-electrode	Detection method	Detection range	Detection limit	Ref.
H_2O_2	Ti_3C_2	Voltammetric	0.1-260 μM	20 nM	[96]
H_2O_2	Ti_3C_2-GO	Voltammetric	2 μM-1mM	1.95 mM	[97]
H_2O_2	$Ti_3C_2T_x$	Chronoampero-metric	N/A	0.7 nM	[98]
H_2O_2	Pt-$Ti_3C_2T_x$	Amperometric	490 μM-53.6mM	448 nM	[99]
H_2O_2	TiO_2-naf-Ti_3C_2	Amperometry	0.1-380 μM	14 nM	[100]

Glucose, Lactate	GO_x/CNTs/ $Ti_3C_2T_x$/PB	Chronoampero-metry	10 μM-1.5 mM, 10 μM-22 mM	0.33 mM, 0.67 mM	[101]
Urea, Uric Acid, Creatinine	MB/$Ti_3C_2T_x$/SPE	Squarewave voltammetry	0.1-3 mM, 30-500 mM, 10-400 μM	0.02 mM, 5 mM, 1.2 μM	[102]
Adrenaline	Ti_2CT_x/GCPE	Chronoampero-metry	0.02-10 mM	9.5 nM	[103]
Dopamine	Ti_3C_2	Conductometric	100 nM-50 mM	100 nM	[104]
Organophosphorus pesticides	$Ti_3C_2T_x$	Amperometric	0.01 pM- 10 nM	0.003 pM	[105]
Malathion	AChE/Ag@ $Ti_3C_2T_x$	Voltammetric	0.01 pM- 0.01 μM	0.00327 pM	[106]
Carbendazim	$Ti_3C_2T_x$	Voltammetric	50 nM-100 mM	10.3 nM	[107]
Methamidophos	AChE/Ti_3C_2/Au/ MnO_2/ Mn_3O_4	Voltammetric	1 pM- 1 μM	0.134 pM	[108]
Single nucleotide, Tripropylamine	$Ti_3C_2T_x$	Electrochemi-luminescent	N/A, 0.1 nM-1 mM	1 nM, 5 nM	[109]
Nitrite	Na/Hb/Ti_3C_2/GCE	Amperometry	0.5-11800 μM	0.12 μM	[110]

N/A = No available data.

5.1 Electrochemical biosensor

After the extensive utilization of several semiconductor nano/microparticles, carbon-based nanomaterials and other metals/non-metals in biosensing [111–116], MXene materials were the most recent to utilized in this area. The morphology, excellent conductivity, good surface chemistry and biocompatibility were utmost prominent features of MXene materials, which made them extremely appropriate materials for biosensors [117–122]. Several electrodes made with MXene materials have been successfully fabricated and effectively applied for immobilization of biological receptor on their surfaces [105,110,117–119,121]. It was difficult to direct electron transfer (DET) among the enzymes and protein because the location of their redox centers were deeply rooted [111]. MXene materials easily permitted the immobilization of protein/enzymes on their surfaces. Hence, they can serve as promising materials for DET with high sensitivity and selectivity, fast electrode kinetics and lower detection limits. Wang et al.

[96] synthesized Ti_3C_2 with the flake-like morphology, and their results suggested that it had a high propensity to be applied organ-like curve monolayers closed from one side and opened at the other for accessible enzyme adsorption as shown in Fig. 6 [100]. In short, the monolayers surface functional groups could easily trap the enzymes and channeled towards the insides of materials. The MXenes presented outstanding charge-transfer capabilities, which made them act as good medias for charge transfer among the protein and electrode made of Ti_3C_2/heomoglobin/nafion and glassy carbon electrode (GCE). The synthesized biosensor also exhibited excellent performance towards the detection of hydrogen peroxide (H_2O_2) with a ranged between 0.1−260 µm.

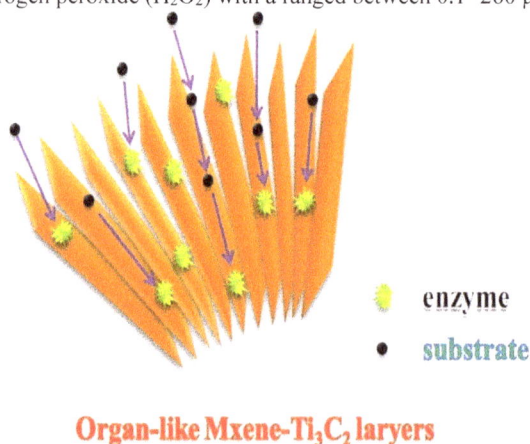

enzyme

substrate

Organ-like Mxene-Ti₃C₂ laryers

Fig. 6 Schematic diagram of the organ-like MXene-Ti₃C₂ structure capturing hemoglobin [100].

Moreover, the enzyme immobilized MXenes working can be further enhanced by increasing the surface area by decorating different metal/metal oxides nanoparticles on the MXene surfaces. The nanoparticles-MXene heterojunctions assisted in combining the enzyme/protein receptor on the nanoparticles-MXene interfaces using a physical adsorption technique. Additionally, the MXene-based biosensors also helped the enzyme/protein to maintain their bioactivities without changing the actual confirmation. Especially, the MXene-based enzymes/protein biosensors were well investigated in the electrochemical determination of various biomolecules. For instance, glucose oxidase (GO_x) was decorated on the surface of nafion-Au-MXene and typically utilized as a biosensor for glucose over GCE [96]. Their results proposed a new route to investigate

Toxic Gas Sensors and Biosensors
Materials Research Foundations **92** (2021) 107-138

Materials Research Forum LLC
https://doi.org/10.21741/9781644901175-4

the diagnostics at an ultra-advanced stage for irregularities of glucose level in blood. As mentioned earlier, it was difficult to accomplish the DET among the pure working electrode and redox centers of GO_x because of the complex GO_x scaffold. In the MXene biosensor, $Ti_3C_2T_x$ exhibited much-improved transfer of electron among the active redox flavin adenine dinucleotide (FAD) of GO_x and electrode interface. The performance of $Ti_3C_2T_x$ sensing was highly enhanced after the decoration of Au, which declined the effect of protein shells covering the FAD.

5.2   Gas sensing

MXenes have also been successfully applied as gas sensors due to the direct charge transfer mechanism [123]. A recent study suggested that monolayers MXene-Ti_2C were extremely good captures and sensors for ammonia (NH_3) gas. Fig. 7 [83] shows that the nitrogen (N) atom presented in NH_3 was situated exactly above the Ti atom of Ti_2CO_2, which demonstrated the higher binding energies than those of the hydrogen (H_2), methane (CH_4), carbon monoxide (CO), carbon dioxide (CO_2), nitrogen (N_2) and nitrogen dioxide (NO_2). The NH_3 gas adsorbed on the Ti_2CO_2 could simply be released through declining the biaxial strains. In the meantime, the electrical conductivity of Ti_2CO_2 was highly improved after the NH_3 adsorption [83].

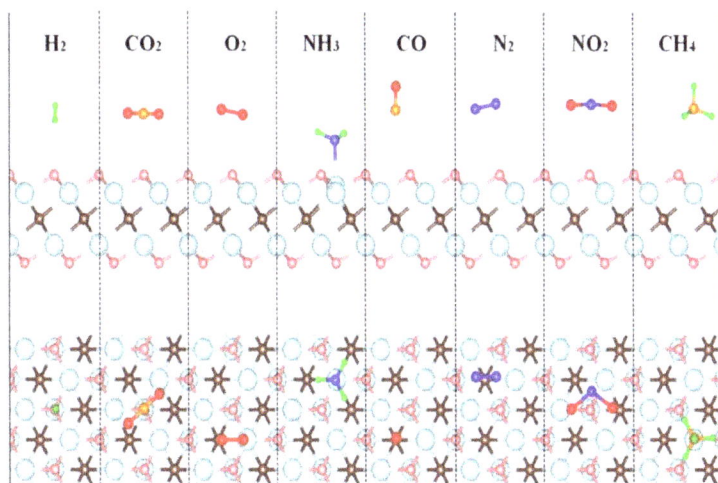

Fig. 7 Schematic diagram of H_2, CH_4, CO, CO_2, NH_3, NO_2, N_2 and O_2 molecules adsorption on Ti_2CO_2 MXene [83].

In an investigation reported by Persson et al. [124], they observed that the Ti_3C_2 presented much higher CO_2 capture affinity than that of the N_2. Additionally, the synthesized Ti_3C_2 sensor also had the capability to detect several other gasses, including acetone, methanol, ethanol and HN_3 at the room temperature. Theoretical studies showed that the selectivity of MXene materials can be enhanced by oxygen functionalization. For instance, Nb_2CO_2 and Ti_2CO_2 have the selectivity towards NH_3 gas and Mo_2CO_2 and V_2CO_2 presented selectivity towards NO gas [125]. Recently, it was predicted that the monolayer Sc_2CO_2 was a good candidate for the sulfur dioxide (SO_2) gas sensor [126]. Among the various investigated MXenes, the Sc_2CO_2 exhibited the highest SO_2 adsorption strength, which can be controlled by applying an external electric field.

5.3 Macromolecule and cell detection

Besides the gases and small molecules, MXenes have vast potential towards the detection of macromolecules and cells. For instance, Xue et al. [127] synthesized three quantum dots MXene (Ti_3C_2 MQD) materials using a hydrothermal synthesis technique. The three different materials were prepared by three different temperatures 100°C, 120°C and 150°C, respectively. These MQDs presented photoluminescence properties with the quantum yield about 10%, and they were also effectively applied as a probe for RAW64.7 cells *in vitro* bio-imaging. Xu et al. [128] synthesized Ti_3C_2-micropattern-based device named Field-Effect Transistor (FET) and first time applied it for the detection of neural activity. The FET device was effectively able to detect neurotransmitters dopamine using doping effect, motivated through *p-p* interaction among the electron of –OH or –F terminal groups and dopamine. Moreover, the FET was also successfully applied as a probe towards hippocampal neurons with outstanding biocompatibility. In a summary, MXene materials have an extremely high potential for the synthesis of sensors with extensive applications in biomedical detection and environmental applications due to their hydrophilic surface, conductivity and biocompatibility.

Conclusion and future prospects

In this chapter, recent research on the synthesis of MXene and MXene-related materials for catalytic applications in gaseous removal and biosensors were systematically reviewed. The MXene-based materials were called for these exceptional applications as they were sparked by the salient characters and properties, such as high metallic conductivity, hardness, hydrophilic nature, tunable band gap structure as well as large surface area. From the viewpoint of preparation and material engineering, most of MXenes have been fabricated till present were applied through selectively etching Al from the MAX phase using the HF as the etchant. Owing to the high HF toxicity, alternative techniques using less hazard or F-free etchants should be devoted as other options for transitional HF. Meanwhile, the synthesis challenge together with the MXene surface chemical functionalization in order to be able to accomplish superior chemical processes in catalysis, adsorption, and sensing. Besides, optimization of fabrication strategies was also a requirement for imperative advancement in the massive amount of production of these MXene materials.

Remarkably, the family of MXene-related materials was a novel and new emerging platform for the advanced catalysis realm both in gaseous removal and biosensing fields. Despite of these extensive works, the working mechanisms and insight catalytic pathways still remained ambiguous and hence, deserved in-depth future studies. In addition, high selectivity, activity, and stability were essential for achieving long-term practical application. Equally important, it is crucial to explicate the role of the MXene as well as the key elementary reaction steps, required further in-depth investigation. This work could be attained by the use of theoretical simulation and laboratory testing together with advanced spectroscopic characterization to comprehensively decipher on the molecular to atomic scale of the catalyst active site. Hence, it is of great importance to bridge the collaboration gap among material scientists, computational engineers with complimentary expertise to discuss and explore theory and experiments for MXene-based catalysts. Furthermore, understanding MXenes toxicity and its environmental impact would substantially amplify their applications in gaseous removal and biosensing areas and pave a potential avenue for the development of MXene-based catalysts fulfilling environmental safety needs. Taking into account on these considerations, the application of innovative MXenes materials in environmental and biosensing areas would certainly show more exciting practical use in the future.

Acknowledgments

The financial supports from Universiti Tunku Abdul Rahman (UTARRF/2019–C1/L03), Ministry of Higher Education of Malaysia (FRGS/1/2016/TK02/UTAR/02/1 and FRGS/1/2019/TK02/UTAR/02/4), Research funds of The Guangxi Key Laboratory of Theory and Technology for Environmental Pollution Control (1801K012 and 1801K013), and special funding for Guangxi "Bagui Scholar" construction project were gratefully acknowledged.

References

[1]   S.M. Lam, M.W. Kee, K.A. Wong, Z.H. Jaffari, H.Y. Chai, J.C. Sin, A newly emerging MXene nanomaterial for environmental applications, MXene: Fund. Appl. 51 (2019) 20–60. http://doi.org/10.21741/9781644900253-2

[2]   J. Zhu, E. Ha, G. Zhao, Y. Zhou, D. Huang, G. Yue, L. Hu, N. Sun, Y. Wang, L. Yoon, S. Lee, C. Xu, K. Wong, D. Astruc, P. Zhao, Recent advance in MXenes: A promising 2D material for catalysis, sensor and chemical adsorption, Coord. Chem. Rev. 352 (2017) 306–327. https://doi.org/10.1016/j.ccr.2017.09.012

[3]   C. Tan, X. Cao, X. Wu, Q. He, J. Yang, X. Zhang, J. Chen, W. Zhao, S. Han, G. Nam, M. Sindoro, H. Zhang, Recent advances in ultrathin two-dimensional nanomaterials, Chem. Rev. 117 (2017) 6225–6331. https://doi.org/10.1021/acs.chemrev.6b00558

[4]   H. Zhang, Ultrathin two-dimensional nanomaterials, ACS Nano. 9 (2015) 9451–9469. https://doi.org/10.1021/acsnano.5b05040

[5]   G.R. Bhimanapati, Z. Lin, V. Meunier, Y. Jung, J. Cha, S. Das, D. Xiao, Y. Son, M.S. Strano, V.R. Cooper, others, Recent advances in two-dimensional materials beyond graphene, ACS Nano. 9 (2015) 11509–11539. https://doi.org/10.1021/acsnano.5b05556

[6]   S.Z. Butler, S.M. Hollen, L. Cao, Y. Cui, J.A. Gupta, H.R. Gutiérrez, T.F. Heinz, S.S. Hong, J. Huang, A.F. Ismach, others, Progress, challenges, and opportunities in two-dimensional materials beyond graphene, ACS Nano. 7 (2013) 2898–2926. https://doi.org/10.1021/nn400280c

[7]   A. Gupta, T. Sakthivel, S. Seal, Recent development in 2D materials beyond graphene, Prog. Mater. Sci. 73 (2015) 44–126. https://doi.org/10.1016/j.pmatsci.2015.02.002

[8]   Z. Lai, Y. Chen, C. Tan, X. Zhang, H. Zhang, Self-assembly of two-dimensional nanosheets into one-dimensional nanostructures, Chem. 1 (2016) 59–77. https://doi.org/10.1016/j.chempr.2016.06.011

[9]   X. Huang, Z. Zeng, H. Zhang, Metal dichalcogenide nanosheets: preparation, properties and applications, Chem. Soc. Rev. 42 (2013) 1934–1946. https://doi.org/10.1039/C2CS35387C

[10]   C. Tan, H. Zhang, Two-dimensional transition metal dichalcogenide nanosheet-based composites, Chem. Soc. Rev. 44 (2015) 2713–2731. https://doi.org/10.1039/C4CS00182F

[11]   R. Lv, J.A. Robinson, R.E. Schaak, D. Sun, Y. Sun, T.E. Mallouk, M. Terrones, Transition metal dichalcogenides and beyond: synthesis, properties, and applications of single-and few-layer nanosheets, Acc. Chem. Res. 48 (2014) 56–64. https://doi.org/10.1021/ar5002846

[12]   Y. Lin, T. V Williams, J.W. Connell, Soluble, exfoliated hexagonal boron nitride nanosheets, J. Phys. Chem. Lett. 1 (2009) 277–283. https://doi.org/10.1021/jz9002108

[13]   Q. Weng, X. Wang, X. Wang, Y. Bando, D. Golberg, Functionalized hexagonal boron nitride nanomaterials: emerging properties and applications, Chem. Soc. Rev. 45 (2016) 3989–4012. https://doi.org/10.1039/C5CS00869G

[14]   L.H. Li, Y. Chen, Atomically thin boron nitride: unique properties and applications, Adv. Funct. Mater. 26 (2016) 2594–2608. https://doi.org/10.1002/adfm.201504606

[15]   H. Liu, Y. Du, Y. Deng, D.Y. Peide, Semiconducting black phosphorus: synthesis, transport properties and electronic applications, Chem. Soc. Rev. 44 (2015) 2732–2743. https://doi.org/10.1039/C4CS00257A

[16]   V. Eswaraiah, Q. Zeng, Y. Long, Z. Liu, Black phosphorus nanosheets: synthesis, characterization and applications, Small. 12 (2016) 3480–3502. https://doi.org/10.1002/smll.201600032

[17]   S.M. Lam, J.C. Sin, A.R. Mohamed, A review on photocatalytic application of g-C_3N_4/semiconductor (CNS) nanocomposites towards the erasure of dyeing wastewater, Mater. Sci. Semicond. Process. 47 (2016) 62–84. https://doi.org/10.1016/j.mssp.2016.02.019

[18]   R. Brec, Review on structural and chemical properties of transition metal phosphorus trisulfides MPS3, in: Intercalation Layer. Mater., Springer, 1986: pp. 93–124. https://doi.org/10.1007/978-1-4757-5556-5_4

[19]   M. Afzaal, P. O'Brien, Recent developments in II-VI and III-VI semiconductors and their applications in solar cells, J. Mater. Chem. 16 (2006) 1597–1602. https://doi.org/10.1039/B512182E

[20]   Q. Wang, D.O. Hare, Recent advances in the synthesis and application of layered double hydroxide (LDH) nanosheets, Chem. Rev. 112 (2012) 4124–4155. https://doi.org/10.1021/cr200434v

[21]   R. Ma, T. Sasaki, Two-dimensional oxide and hydroxide nanosheets: controllable high-quality exfoliation, molecular assembly, and exploration of functionality, Acc. Chem. Res. 48 (2014) 136–143. https://doi.org/10.1021/ar500311w

[22]   J.W. Colson, A.R. Woll, A. Mukherjee, M.P. Levendorf, E.L. Spitler, V.B. Shields, M.G. Spencer, J. Park, W.R. Dichtel, Oriented 2D covalent organic framework thin films on single-layer graphene, Science. 332 (2011) 228–231. https://doi.org/10.1126/science.1202747

[23]   S.L. Cai, W.G. Zhang, R.N. Zuckermann, Z.-T. Li, X. Zhao, Y. Liu, The organic flatland-recent advances in synthetic 2D organic layers, Adv. Mater. 27 (2015) 5762–5770. https://doi.org/10.1002/adma.201500124

[24]   B. Li, H.-M. Wen, W. Zhou, J.Q. Xu, B. Chen, Porous metal-organic frameworks: promising materials for methane storage, Chem. 1 (2016) 557–580. https://doi.org/10.1016/j.chempr.2016.09.009

[25]   C. Tan, X. Qi, X. Huang, J. Yang, B. Zheng, Z. An, R. Chen, J. Wei, B.Z. Tang, W. Huang, Single-layer transition metal dichalcogenide nanosheet-assisted assembly of aggregation-induced emission molecules to form organic nanosheets with enhanced fluorescence, Adv. Mater. 26 (2014) 1735–1739. https://doi.org/10.1002/adma.201304562

[26]   Z. Fan, X. Huang, C. Tan, H. Zhang, Thin metal nanostructures: synthesis, properties and applications, Chem. Sci. 6 (2015) 95–111. https://doi.org/10.1039/C4SC02571G

[27]   Z.H. Jaffari, S. Lam, J. Sin, H. Zeng, Boosting visible light photocatalytic and antibacterial performance by decoration of silver on magnetic spindle-like bismuth ferrite, Mater. Sci. Semicond. Process. 101 (2019) 103–115. https://doi.org/10.1016/j.mssp.2019.05.036

[28]   J. Song, L. Xu, J. Li, J. Xue, Y. Dong, X. Li, H. Zeng, Monolayer and few-layer all-inorganic perovskites as a new family of two-dimensional semiconductors for printable optoelectronic devices, Adv. Mater. 28 (2016) 4861–4869. https://doi.org/10.1002/adma.201600225

[29]   Y. Ebina, T. Sasaki, M. Watanabe, Study on exfoliation of layered perovskite-type niobates, Solid State Ionics. 151 (2002) 177–182. https://doi.org/10.1016/S0167-2738(02)00707-5

[30]   M. Hargittai, Molecular structure of metal halides, Chem. Rev. 100 (2000) 2233–2302. https://doi.org/10.1021/cr970115u

[31]   Q. Lu, M. Zhao, J. Chen, B. Chen, C. Tan, X. Zhang, Y. Huang, J. Yang, F. Cao, Y. Yu, others, In situ synthesis of metal sulfide nanoparticles based on 2D metal-

Materials Research Forum LLC
https://doi.org/10.21741/9781644901175-4

organic framework nanosheets, Small. 12 (2016) 4669–4674.
https://doi.org/10.1002/smll.201600976

[32]   Y. Peng, Y. Li, Y. Ban, H. Jin, W. Jiao, X. Liu, W. Yang, Metal-organic
framework nanosheets as building blocks for molecular sieving membranes, Science.
346 (2014) 1356–1359. https://doi.org/10.1126/science.1254227

[33]   M. Armand, L. Coic, P. Palvadeau, J. Rouxel, The M-A-X transition metal
oxyhalides: A new class of lamellar cathode material, J. Power Sources. 3 (1978) 137–
144. https://doi.org/10.1016/0378-7753(78)80012-3

[34]   Y. Li, W. Shen, Morphology-dependent nanocatalysts: rod-shaped oxides, Chem.
Soc. Rev. 43 (2014) 1543–1574. https://doi.org/10.1039/C3CS60296F

[35]   T. Selvam, A. Inayat, W. Schwieger, Reactivity and applications of layered
silicates and layered double hydroxides, Dalton. Trans. 43 (2014) 10365−10387.
https://doi.org/10.1039/C4DT00573B

[36]   C. Tan, H. Zhang, Wet-chemical synthesis and applications of non-layer structured
two-dimensional nanomaterials, Nat. Commun. 6 (2015) 7873.
https://doi.org/10.1038/ncomms8873

[37]   M. Naguib, V.N. Mochalin, M.W. Barsoum, Y. Gogotsi, 25th anniversary article:
MXenes: a new family of two-dimensional materials, Adv. Mater. 26 (2014) 992–
1005. https://doi.org/10.1002/adma.201304138

[38]   B. Anasori, M.R. Lukatskaya, Y. Gogotsi, 2D metal carbides and nitrides
(MXenes) for energy storage, Nat. Rev. Mater. 2 (2017) 16098.
https://doi.org/10.1038/natrevmats.2016.98

[39]   M. Khazaei, M. Arai, T. Sasaki, C.Y. Chung, N.S. Venkataramanan, M. Estili, Y.
Sakka, Y. Kawazoe, Novel electronic and magnetic properties of two-dimensional
transition metal carbides and nitrides, Adv. Funct. Mater. 23 (2013) 2185–2192.
https://doi.org/10.1002/adfm.201202502

[40]   Y. Gao, L. Wang, A. Zhou, Z. Li, J. Chen, H. Bala, Q. Hu, X. Cao, Hydrothermal
synthesis of TiO_2/Ti_3C_2 nanocomposites with enhanced photocatalytic activity, Mater.
Lett. 150 (2015) 62–64. https://doi.org/10.1016/j.matlet.2015.02.135

[41]   X. Zhang, J. Lei, D. Wu, X. Zhao, Y. Jing, Z. Zhou, A Ti-anchored Ti_2CO_2
monolayer (MXene) as a single-atom catalyst for CO oxidation, J. Mater. Chem. A. 4
(2016) 4871–4876. https://doi.org/10.1039/C6TA00554C

[42]   M. Naguib, M. Kurtoglu, V. Presser, J. Lu, J. Niu, M. Heon, L. Hultman, Y. Gogotsi, M.W. Barsoum, Two-dimensional nanocrystals produced by exfoliation of Ti$_3$AlC$_2$, Adv. Mater. 23 (2011) 4248–4253. https://doi.org/10.1002/adma.201102306

[43]   M. Naguib, J. Halim, J. Lu, K.M. Cook, L. Hultman, Y. Gogotsi, M.W. Barsoum, New two-dimensional niobium and vanadium carbides as promising materials for Li-ion batteries, J. Am. Chem. Soc. 135 (2013) 15966–15969. https://doi.org/10.1021/ja405735d

[44]   M. Naguib, O. Mashtalir, J. Carle, V. Presser, J. Lu, L. Hultman, Y. Gogotsi, M.W. Barsoum, Two-dimensional transition metal carbides, ACS Nano. 6 (2012) 1322–1331. https://doi.org/10.1021/nn204153h

[45]   J. Halim, S. Kota, M.R. Lukatskaya, M. Naguib, M.Q. Zhao, E.J. Moon, J. Pitock, J. Nanda, S.J. May, Y. Gogotsi, M.W. Barsoum, Synthesis and characterization of 2D molybdenum carbide (MXene), Adv. Funct. Mater. 26 (2016) 3118–3127. https://doi.org/10.1002/adfm.201505328

[46]   Z.W. Seh, K.D. Fredrickson, B. Anasori, J. Kibsgaard, A.L. Strickler, M.R. Lukatskaya, Y. Gogotsi, T.F. Jaramillo, A. Vojvodic, Two-dimensional molybdenum carbide (MXene) as an efficient electrocatalyst for hydrogen evolution, ACS Energy Lett. 1 (2016) 589–594. https://doi.org/10.1021/acsenergylett.6b00247

[47]   J. Yang, M. Naguib, M. Ghidiu, L.-M. Pan, J. Gu, J. Nanda, J. Halim, Y. Gogotsi, M.W. Barsoum, Two-Dimensional Nb-Based M$_4$C$_3$ Solid Solutions (MXenes), J. Am. Ceram. Soc. 99 (2016) 660–666. https://doi.org/10.1111/jace.13922

[48]   B. Anasori, Y. Xie, M. Beidaghi, J. Lu, B.C. Hosler, L. Hultman, P.R.C. Kent, Y. Gogotsi, M.W. Barsoum, Two-dimensional, ordered, double transition metals carbides (MXenes), ACS Nano. 9 (2015) 9507–9516. https://doi.org/10.1021/acsnano.5b03591

[49]   J. Zhou, X. Zha, F.Y. Chen, Q. Ye, P. Eklund, S. Du, Q. Huang, A two-dimensional zirconium carbide by Selective etching of Al$_3$C$_3$ from nanolaminated Zr$_3$Al$_3$C$_5$, Angew. Chemie Int. Ed. 55 (2016) 5008–5013. https://doi.org/10.1002/anie.201510432

[50]   J. Zhou, X. Zha, X. Zhou, F. Chen, G. Gao, S. Wang, C. Shen, T. Chen, C. Zhi, P. Eklund, others, Synthesis and electrochemical properties of two-dimensional hafnium carbide, ACS Nano. 11 (2017) 3841–3850. https://doi.org/10.1021/acsnano.7b00030

[51]   M. Ghidiu, M.R. Lukatskaya, M.-Q. Zhao, Y. Gogotsi, M.W. Barsoum, Conductive two-dimensional titanium carbide'clay'with high volumetric capacitance, Nature. 516 (2014) 78. https://doi.org/10.1038/nature13970

[52]   A. Shahzad, K. Rasool, M. Nawaz, W. Miran, J. Jang, M. Moztahida, K.A. Mahmoud, D.S. Lee, Heterostructural $TiO_2/Ti_3C_2T_x$ (MXene) for photocatalytic degradation of antiepileptic drug carbamazepine, Chem. Eng. J. 349 (2018) 748–755. https://doi.org/10.1016/j.cej.2018.05.148

[53]   A. Lipatov, M. Alhabeb, M.R. Lukatskaya, A. Boson, Y. Gogotsi, A. Sinitskii, Effect of synthesis on quality, electronic properties and environmental stability of individual monolayer Ti_3C_2 MXene flakes, Adv. Electron. Mater. 2 (2016) 1600255. https://doi.org/10.1002/aelm.201600255

[54]   L. Wang, H. Zhang, B. Wang, C. Shen, C. Zhang, Q. Hu, A. Zhou, B. Liu, Synthesis and electrochemical performance of $Ti_3C_2T_x$ with hydrothermal process, Electron. Mater. Lett. 12 (2016) 702–710. https://doi.org/10.1007/s13391-016-6088-z

[55]   A. Feng, Y. Yu, Y. Wang, F. Jiang, Y. Yu, L. Mi, L. Song, Two-dimensional MXene Ti_3C_2 produced by exfoliation of Ti_3AlC_2, Mater. Des. 114 (2017) 161–166. https://doi.org/10.1016/j.matdes.2016.10.053

[56]   P. Urbankowski, B. Anasori, T. Makaryan, D. Er, S. Kota, P.L. Walsh, M. Zhao, V.B. Shenoy, M.W. Barsoum, Y. Gogotsi, Synthesis of two-dimensional titanium nitride Ti_4N_3 (MXene), Nanoscale. 8 (2016) 11385–11391. https://doi.org/10.1039/C6NR02253G

[57]   C. Xu, L. Wang, Z. Liu, L. Chen, J. Guo, N. Kang, X.-L. Ma, H.-M. Cheng, W. Ren, Large-area high-quality 2D ultrathin Mo_2C superconducting crystals, Nat. Mater. 14 (2015) 1135. https://doi.org/10.1038/nmat4374

[58]   M.R. Lukatskaya, O. Mashtalir, C.E. Ren, Y. Dall'Agnese, P. Rozier, P.L. Taberna, M. Naguib, P. Simon, M.W. Barsoum, Y. Gogotsi, Cation intercalation and high volumetric capacitance of two-dimensional titanium carbide, Science. 341 (2013) 1502–1505. https://doi.org/10.1126/science.1241488

[59]   J. Luo, W. Zhang, H. Yuan, C. Jin, L. Zhang, H. Huang, C. Liang, Y. Xia, J. Zhang, Y. Gan, others, Pillared structure design of MXene with ultralarge interlayer spacing for high-performance lithium-ion capacitors, ACS Nano. 11 (2017) 2459–2469. https://doi.org/10.1021/acsnano.6b07668

[60]   O. Mashtalir, M. Naguib, V.N. Mochalin, Y. Dall'Agnese, M. Heon, M.W. Barsoum, Y. Gogotsi, Intercalation and delamination of layered carbides and carbonitrides, Nat. Commun. 4 (2013) 1716. https://doi.org/10.1038/ncomms2664

[61]   J. Liu, Y. Liu, D. Xu, Y. Zhu, W. Peng, Y. Li, F. Zhang, X. Fan, Hierarchical "nanoroll" like $MoS_2/Ti_3C_2T_x$ hybrid with high electrocatalytic hydrogen evolution

Toxic Gas Sensors and Biosensors   Materials Research Forum LLC
Materials Research Foundations **92** (2021) 107-138   https://doi.org/10.21741/9781644901175-4

activity, Appl. Catal. B Environ. 241 (2019) 89–94.
https://doi.org/10.1016/j.apcatb.2018.08.083

[62]   M. Naguib, R.R. Unocic, B.L. Armstrong, J. Nanda, Large-scale delamination of
multi-layers transition metal carbides and carbonitrides "MXenes", Dalt. Trans. 44
(2015) 9353–9358. https://doi.org/10.1039/C5DT01247C

[63]   A. Vaughn, J. Ball, T. Heil, D.J. Morgan, G.I. Lampronti, G. Maršalkait, C.L.
Raston, N.P. Power, S. Kellici, Selective calixarene-directed synthesis of MXene
plates, crumpled sheets, spheres, and scrolls, Chem. Eur. J. 23 (2017) 8128–8133.
https://doi.org/10.1002/chem.201701702

[64]   I.R. Shein, A.L. Ivanovskii, Graphene-like nanocarbides and nanonitrides of d
metals (MXenes): synthesis, properties and simulation, Micro Nano Lett. 8 (2013) 59–
62. https://doi.org/10.1049/mnl.2012.0797

[65]   N. Bovenzi, M. Breitkreiz, P. Baireuther, T.E. O'Brien, J. Tworzydło, Inanç
Adagideli, C.W.J. Beenakker, Chirality blockade of Andreev reflection in a magnetic
Weyl semimetal, Phys. Rev. B. 96 (2017) 35437.
https://doi.org/10.1103/PhysRevB.96.035437

[66]   G.R. Berdiyorov, Effect of surface functionalization on the electronic transport
properties of Ti_3C_2 MXene, A Lett. J. Explor. Front. Phys. 111 (2015) 67002.
https://doi.org/10.1209/0295-5075/111/67002

[67]   C. Chen, X. Ji, K. Xu, B. Zhang, L. Miao, J. Jiang, Prediction of T- and H-phase
two-dimensional transition-metal carbides/nitrides and their semiconducting – metallic
phase transition, ChemPhysChem. 18 (2017) 1897–1902.
https://doi.org/10.1002/cphc.201700111

[68]   G. Liu, W. Jin, N. Xu, Two-dimensional-material membranes: A new family of
high-performance separation membranes, Angew. Chem. Int. Ed. 55 (2016) 13384-
13397. https://doi.org/10.1002/anie.201600438

[69]   J. Zhu, J. Hou, A. Uliana, Y. Zhang, M. Tian, B. Van Der Bruggen, The rapid
emergence of two-dimensional nanomaterials for high-performance separation
membranes, J. Mater. Chem. A 6 (2018) 3773-3792.
https://doi.org/10.1039/C7TA10814A

[70]   X. Gao, Z.K. Li, J. Xue, Y. Qian, L.Z. Zhang, J. Caro, H. Wang, Titanium carbide
$Ti_3C_2T_x$ (MXene) enhanced PAN nanofiber membrane for air purification, J. Membr.
Sci. 586 (2019) 162-169. https://doi.org/10.1016/j.memsci.2019.05.058

[71]   R. Han, Y. Xie, X. Ma, Crosslinked P84 copolyimide/MXene mixed matrix membrane with excellent solvent resistance and perm selectivity, Chinese J. Chem. Eng. 27 (2019) 877-883. https://doi.org/10.1016/j.cjche.2018.10.005

[72]   Z. Xu, G. Liu, H. Ye, W. Jin, Z. Cui, Two-dimensional MXene incorporated chitosan mixed-matrix membranes for efficient solvent dehydration, J. Membr. Sci. 563 (2018) 625-632. https://doi.org/10.1016/j.memsci.2018.05.044

[73] L. Ding, Y. Wei, L. Li, T. Zhang, H. Wang, J. Xue, L.-X. Ding, S. Wang, J. Caro, Y. Gogotsi, MXene molecular sieving membranes for highly efficient gas separation, Nat. Commun. 9 (2018) 155. https://doi.org/10.1038/s41467-017-02529-6

[74] B.-M. Jun, S. Kim, J. Heo, C.M. Park, N. Her, M. Jang, Y. Huang, J. Han, Y. Yoon, Review of MXenes as new nanomaterials for energy storage/delivery and selected environmental applications, Nano Res. 12 (2019) 471-487. https://doi.org/10.1007/s12274-018-2225-3

[75]   A. Lipatov, H. Lu, M. Alhabeb, B. Anasori, A. Gruverman, Y. Gogotsi, A. Sinitskii, Elastic properties of 2D $Ti_3C_2T_x$ MXene monolayers and bilayers, Sci. Adv.s 4 (2018) eaat0491. https://doi.org/10.1126/sciadv.aat0491

[76]   Y. Jin, Y. Fan, X. Meng, W. Zhang, B. Meng, N. Yang, S. Liu, Theoretical and experimental insights into the mechanism for gas separation through nanochannels in 2D laminar MXene membranes, Processes 7 (2019) 751. https://doi.org/10.3390/pr7100751

[77]   L. Ding, Y. Wei, Y. Wang, H. Chen, J. Caro, H. Wang, A Two-dimensional lamellar membrane: MXene nanosheet stacks, Angew. Chem. Int. Ed. 56 (2017) 1825-1829. https://doi.org/10.1002/anie.201609306

[78]   Y. Fan, L. Wei, X. Meng, W. Zhang, N. Yang, Y. Jin, X. Wang, M. Zhao, S. Liu, An unprecedented high-temperature-tolerance 2D laminar MXene membrane for ultrafast hydrogen sieving, J. Membr. Sci. 569 (2019) 117-123. https://doi.org/10.1016/j.memsci.2018.10.017

[79]   J. Shen, G. Liu, Y. Ji, Q. Liu, L. Cheng, K. Guan, M. Zhang, G. Liu, J. Xiong, J. Yang, W. Jin, 2D MXene nanofilms with tunable gas transport channels, Adv. Funct. Mater. 28 (2018) 1801511. https://doi.org/10.1002/adfm.201801511

[80]   L. Wang, M.S.H. Boutilier, P.R. Kidambi, D. Jang, N.G. Hadjiconstantinou, R. Karnik, Fundamental transport mechanisms, fabrication and potential applications of nanoporous atomically thin membranes, Nat. Nanotechnol. 12 (2017) 509-522. https://doi.org/10.1038/nnano.2017.72

[81]   Y. Zhang, L. Wang, N. Zhang, Z. Zhou, Adsorptive environmental applications of MXene nanomaterials: A review, RSC Adv. 8 (2018) 19895-19905. https://doi.org/10.1039/C8RA03077D

[82]   S.J. Kim, H.J. Koh, C.E. Ren, O. Kwon, K. Maleski, S.Y. Cho, B. Anasori, C.K. Kim, Y.K. Choi, J. Kim, Y. Gogotsi, H.T. Jung, Metallic $Ti_3C_2T_x$ MXene gas sensors with ultrahigh signal-to-noise ratio, ACS Nano 12 (2018) 986-993. https://doi.org/10.1021/acsnano.7b07460

[83]   X.F. Yu, Y.C. Li, J.B. Cheng, Z.B. Liu, Q.Z. Li, W.Z. Li, X. Yang, B. Xiao, Monolayer Ti_2CO_2: A promising candidate for NH_3 sensor or capturer with high sensitivity and selectivity, ACS Appl. Mater. Inter. 7 (2015) 13707-13713. https://doi.org/10.1021/acsami.5b03737

[84]   B. Wang, A. Zhou, F. Liu, J. Cao, L. Wang, Q. Hu, Carbon dioxide adsorption of two-dimensional carbide MXenes, J. Adv. Ceram. 7 (2018) 237-245. https://doi.org/10.1007/s40145-018-0275-3

[85]   F. Liu, A. Zhou, J. Chen, H. Zhang, J. Cao, L. Wang, Q. Hu, Preparation and methane adsorption of two-dimensional carbide Ti_2C, Adsorption 22 (2016) 915-922. https://doi.org/10.1007/s10450-016-9795-8

[86]   F. Liu, A. Zhou, J. Chen, J. Jia, W. Zhou, L. Wang, Q. Hu, Preparation of Ti_3C_2 and Ti_2C MXenes by fluoride salts etching and methane adsorptive properties, Appl. Surf. Sci. 416 (2017) 781-789. https://doi.org/10.1016/j.apsusc.2017.04.239

[87]   Q. Hu, D. Sun, Q. Wu, H. Wang, L. Wang, B. Liu, A. Zhou, J. He, MXene: A new family of promising hydrogen storage medium, J. Phys. Chem. A 117 (2013) 14253-14260. https://doi.org/10.1021/jp409585v

[88]   Q. Hu, H. Wang, Q. Wu, X. Ye, A. Zhou, D. Sun, L. Wang, B. Liu, J. He, Two-dimensional Sc_2C: A reversible and high-capacity hydrogen storage material predicted by first-principles calculations, Int. J. Hydrogen Energy 39 (2014) 10606-10612. https://doi.org/10.1016/j.ijhydene.2014.05.037

[89   A. Yadav, A. Dashora, N. Patel, A. Miotello, M. Press, D.C. Kothari, Study of 2D MXene Cr_2C material for hydrogen storage using density functional theory, Appl. Surf. Sci. 389 (2016) 88-95. https://doi.org/10.1016/j.apsusc.2016.07.083

[90]   Y. Li, Y. Guo, W. Chen, Z. Jiao, S. Ma, Reversible hydrogen storage behaviors of Ti_2N MXenes predicted by first-principles calculations, J. Mater. Sci. 54 (2019) 493-505. https://doi.org/10.1007/s10853-018-2854-7

[91] B. Xiao, Y.C. Li, X.F. Yu, J.B. Cheng, MXenes: Reusable materials for NH_3 sensor or capturer by controlling the charge injection, Sensor Actuat. B: Chem. 235 (2016) 103-109. https://doi.org/10.1016/j.snb.2016.05.062

[92] E. Lee, A.V. Mohammadi, Y.S. Yoon, M. Beidaghi, D.J. Kim, Two-dimensional vanadium carbide MXene for gas sensors with ultrahigh sensitivity toward nonpolar gases, ACS Sensor. 4 (2019) 1603-1611. https://doi.org/10.1021/acssensors.9b00303

[93] J. Ren, M. Antonietti, T.-P. Fellinger, Efficient water splitting using a simple Ni/N/C paper electrocatalyst, Adv. Energy Mater. 5 (2015) 1401660. https://doi.org/10.1002/aenm.201401660

[94] T.Y. Ma, J.L. Cao, M. Jaroniec, S.Z. Qiao, Interacting carbon nitride and titanium carbide nanosheets for high-performance oxygen evolution, Angew. Chem. Int. Ed. 55 (2016) 1138–1142. https://doi.org/10.1002/anie.201509758

[95] G.M. Huang, S.Z. Li, L.Z. Liu, L.F. Zhu, Q. Wang, Ti_3C_2 MXene-modified Bi_2WO_6 nanoplates for efficient photodegradation of volatile organic compounds. Appl. Surf. Sci. 503 (2020) 144183–144191. https://doi.org/10.1016/j.apsusc.2019.144183

[96] F. Wang, C. Yang, C. Duan, D. Xiao, Y. Tang, J. Zhu, An organ-like titanium carbide material (MXene) with multilayer structure encapsulating hemoglobin for a mediator-free biosensor, J. Electrochem. Soc. 162 (2015) 16–21. https://doi.org/10.1149/2.0371501jes

[97] J. Zheng, J. Diao, Y. Jin, A. Ding, B. Wang, L. Wu, B. Weng, J. Chen, An inkjet printed Ti_3C_2-GO electrode for the electrochemical sensing of hydrogen peroxide, J. Electrochem. Soc. 165 (2018) 227-231. https://doi.org/10.1016/j.snb.2016.05.062

[98] L. Lorencova, T. Bertok, E. Dosekova, A. Holazova, D. Paprckova, A. Vikartovska, V. Sasinkova, J. Filip, P. Kasak, M. Jerigova, D. Velic, K.A. Mahmoud, J. Tkac, Electrochemical performance of $Ti_3C_2T_x$ MXene in aqueous media: towards ultrasensitive H_2O_2 sensing, Electrochim. Acta. 235 (2017) 471–479. https://doi.org/10.1016/j.electacta.2017.03.073

[99] L. Lorencová, T. Bertok, J. Filip, M. Jerigová, D. Velic, P. Kasák, K.A. Mahmoud, J. Tkac, Highly stable $Ti_3C_2T_x$ (MXene)/Pt nanoparticles-modified glassy carbon electrode for H_2O_2 and small molecules sensing applications, Sens. Actuators B Chem. 263 (2018) 360–368. https://doi.org/10.1016/j.snb.2018.02.124

[100] F. Wang, C. Yang, M. Duan, Y. Tang, J. Zhu, TiO_2 nanoparticle modified organ-like Ti_3C_2 MXene nanocomposite encapsulating hemoglobin for a mediator-free

biosensor with excellent performances, Biosens. Bioelectron. 74 (2015) 1022–1028. https://doi.org/10.1016/j.bios.2015.08.004

[101] Y. Lei, W. Zhao, Y. Zhang, Q. Jiang, J. He, A.J. Baeumner, O.S. Wolfbeis, Z.L. Wang, K.N. Salama, H.N. Alshareef, A MXene-based wearable biosensor system for high-performance in vitro perspiration analysis, Small. 15 (2019) 1901190. https://doi.org/10.1002/smll.201901190

[102] J. Liu, X. Jiang, R. Zhang, Y. Zhang, L. Wu, W. Lu, J. Li, MXene-enabled electrochemical microfluidic biosensor: applications toward multicomponent continuous monitoring in whole blood, Adv. Funct. Mater. 29 (2018) 1807326. https://doi.org/10.1002/adfm.201807326

[103] S.S. Shankar, R.M. Shereema, R.B. Rakhi, Electrochemical determination of adrenaline using MXene/graphite composite paste electrodes, ACS Appl. Mater. Interfaces. 10 (2018) 43343–43351. https://doi.org/10.1021/acsami.8b11741

[104] B. Xu, M. Zhu, W. Zhang, X. Zhen, Z. Pei, Q. Xue, C. Zhi, Ultrathin MXene-micropattern-based field-effect transistor for probing neural activity, Adv. Mater. 28 (2016) 3333–3339. https://doi.org/10.1002/adma.201504657

[105] L. Zhou, X. Zhang, J. Gao, Y. Jiang, X. Zhang, J. Gao, Acetylcholinesterase/chitosan-transition metal carbides nanocomposites-based biosensor for the organophosphate pesticides detection, Biochem. Eng. J. 128 (2017) 243–249. https://doi.org/10.1016/j.bej.2017.10.008

[106] Y. Jiang, X. Zhang, L. Pei, S. Yue, L. Ma, L. Zhou, Z. Huang, Silver nanoparticles modified two-dimensional transition metal carbides as nanocarriers to fabricate acetycholinesterase-based electrochemical biosensor, Chem. Eng. J. 339 (2018) 547–556. https://doi.org/10.1016/j.cej.2018.01.111

[107] D. Wu, M. Wu, J. Yang, H. Zhang, K. Xie, C. Lin, A. Yu, J. Yu, L. Fu, Delaminated $Ti_3C_2T_x$ (MXene) for electrochemical carbendazim sensing, Mater. Lett. 236 (2019) 412–415. https://doi.org/10.1016/j.matlet.2018.10.150

[108] D. Song, X. Jiang, Y. Li, X. Lu, S. Luan, Y. Wang, Y. Li, F. Gao, Metal–organic frameworks-derived MnO_2/Mn_3O_4 microcuboids with hierarchically ordered nanosheets and Ti_3C_2 MXene/ Au NPs composites for electrochemical pesticide detection, J. Hazard. Mater. 373 (2019) 367–376. https://doi.org/10.1016/j.jhazmat.2019.03.083

[109] Y. Fang, X. Yang, T. Chen, G. Xu, M. Liu, J. Liu, Y. Xu, Two-dimensional titanium carbide (MXene)-based solid-state electrochemiluminescent sensor for label-

free single-nucleotide mismatch discrimination in human urine, Sensors Actuators B. Chem. 263 (2018) 400–407. https://doi.org/10.1016/j.snb.2018.02.102

[110] H. Liu, C. Duan, C. Yang, W. Shen, F. Wang, Z. Zhu, A novel nitrite biosensor based on the direct electrochemistry of hemoglobin immobilized on MXene-Ti$_3$C$_2$, Sens. Actuators B Chem., 218 (2015) 60–66. https://doi.org/10.1016/j.snb.2015.04.090

[111] R.B. Rakhi, P. Nayak, C. Xia, H.N. Alshareef, Novel amperometric glucose biosensor based on MXene nanocomposite, Sci. Rep. 6 (2016) 36422. https://doi.org/10.1038/srep36422

[112] G. Liu, W. Jin, N. Xu, Two-dimensional-material membranes: A new family of high-performance separation membranes, Angew. Chem. Int. Ed. 55 (2016) 13384-13397. https://doi.org/10.1002/anie.201600438

[113] J. Zhu, J. Hou, A. Uliana, Y. Zhang, M. Tian, B. Van Der Bruggen, The rapid emergence of two-dimensional nanomaterials for high-performance separation membranes, J. Mater. Chem. A 6 (2018) 3773-3792. https://doi.org/10.1039/C7TA10814A

[114] X. Gao, Z.K. Li, J. Xue, Y. Qian, L.Z. Zhang, J. Caro, H. Wang, Titanium carbide Ti$_3$C$_2$T$_x$ (MXene) enhanced PAN nanofiber membrane for air purification, J. Membr. Sci. 586 (2019) 162-169. https://doi.org/10.1016/j.memsci.2019.05.058

[115] R. Han, Y. Xie, X. Ma, Crosslinked P84 copolyimide/MXene mixed matrix membrane with excellent solvent resistance and perm selectivity, Chinese J. Chem. Eng. 27 (2019) 877-883. https://doi.org/10.1016/j.cjche.2018.10.005

[116] Z. Xu, G. Liu, H. Ye, W. Jin, Z. Cui, Two-dimensional MXene incorporated chitosan mixed-matrix membranes for efficient solvent dehydration, J. Membr. Sci. 563 (2018) 625-632. https://doi.org/10.1016/j.memsci.2018.05.044

[117] L. Ding, Y. Wei, L. Li, T. Zhang, H. Wang, J. Xue, L.-X. Ding, S. Wang, J. Caro, Y. Gogotsi, MXene molecular sieving membranes for highly efficient gas separation, Nat. Commun. 9 (2018) 155. https://doi.org/10.1038/s41467-017-02529-6

[118] B.M. Jun, S. Kim, J. Heo, C.M. Park, N. Her, M. Jang, Y. Huang, J. Han, Y. Yoon, Review of MXenes as new nanomaterials for energy storage/delivery and selected environmental applications, Nano Res. 12 (2019) 471-487. https://doi.org/10.1007/s12274-018-2225-3

[119] A. Lipatov, H. Lu, M. Alhabeb, B. Anasori, A. Gruverman, Y. Gogotsi, A. Sinitskii, Elastic properties of 2D $Ti_3C_2T_x$ MXene monolayers and bilayers, Sci. Adv. 4 (2018) eaat0491. https://doi.org/10.1126/sciadv.aat0491

[120] Y. Jin, Y. Fan, X. Meng, W. Zhang, B. Meng, N. Yang, S. Liu, Theoretical and experimental insights into the mechanism for gas separation through nanochannels in 2D laminar MXene membranes, Processes 7 (2019) 751. https://doi.org/10.3390/pr7100751

[121] L. Ding, Y. Wei, Y. Wang, H. Chen, J. Caro, H. Wang, A Two-dimensional lamellar membrane: MXene nanosheet stacks, Angew. Chem. Int. Ed. 56 (2017) 1825-1829. https://doi.org/10.1002/anie.201609306

[122] Y. Fan, L. Wei, X. Meng, W. Zhang, N. Yang, Y. Jin, X. Wang, M. Zhao, S. Liu, An unprecedented high-temperature-tolerance 2D laminar MXene membrane for ultrafast hydrogen sieving, J. Membr. Sci. 569 (2019) 117-123. https://doi.org/10.1016/j.memsci.2018.10.017

[123] M. Khazaei, A. Ranjbar, M. Ghorbani-asl, M. Arai, T. Sasaki, Y. Liang, S. Yunoki, Nearly free electron states in MXenes, Phys. Rev. B. 93 (2016) 205125. https://doi.org/10.1103/PhysRevB.93.205125

[124] I. Persson, J. Halim, H. Lind, T.W. Hansen, J.B. Wagner, L.-ake Näslund, V. Darakchieva, J. Palisaitis, J. Rosen, P.O.A. Persson, 2D transition metal carbides (MXenes) for carbon capture, Adv. Mater. 31 (2018) 1805472. https://doi.org/10.1002/adma.201805472

[125] A. Junkaew, R. Arroyave, Enhancement of the selectivity of MXenes (M_2C, M = Ti, V, Nb, Mo) via oxygen-functionalization: promising materials for gas sensing and-separation, Phys. Chem. Chem. Phys. 20 (2018) 6073–6082. https://doi.org/10.1039/C7CP08622A

[126] S. Ma, D. Yuan, Z. Jiao, T. Wang, X. Dai, Monolayer Sc_2CO_2: A promising candidate as SO_2 gas sensor or capturer, J. Phys. Chem. C. 121 (2017) 24077–24084. https://doi.org/10.1021/acs.jpcc.7b07921

[127] Q. Xue, H. Zhang, M. Zhu, Z. Pei, H. Li, Z. Wang, Y. Huang, Photoluminescent Ti_3C_2 MXene quantum dots for multicolor cellular imaging, Adv. Mater. 29 (2017) 1604847. https://doi.org/10.1002/adma.201604847

[128] B. Xu, M. Zhu, W. Zhang, X. Zhen, Z. Pei, Q. Xue, C. Zhi, Ultrathin MXene-micropattern-based field-effect transistor for probing neural activity, Adv. Mater. 28 (2016) 3333–3339. https://doi.org/10.1002/adma.201504657

Toxic Gas Sensors and Biosensors   Materials Research Forum LLC
Materials Research Foundations **92** (2021) 139-156   https://doi.org/10.21741/9781644901175-5

Chapter 5

Hydrazine Sensing Technologies

Muhammad Inam Khan[1], Muhammad Tayyab[2], Muhammad Mudassir Hassan[2],
Nawshad Muhammad[2], Awais Ahmad[3], Muhammad Tariq[4], Abdur Rahim[2]*

[1]Department of Physics,COMSATS University Islamabad, Lahore Campus, Defense Road, Off Raiwind Road, Lahore, Pakistan

[2]Interdisciplinary Research Centre in Biomedical Materials (IRCBM), COMSATS University Islamabad, Lahore Campus, Defense Road, Off Raiwind Road, Lahore, Pakistan

[3]Department of Applied Chemistry, Government College University Faisalabad, 38000 Pakistan

[4]National Centre of Excellence in Physical Chemistry, University of Peshawar, KPK, Pakistan

*rahimkhan533@gmail.com

Abstract

Recently, the design, fabrication, and development of the different types of sensing techniques have been reported. Hydrazine is used in many industries such as agriculture, power generation, pharmaceutical, aerospace, and chemical industries. Hydrazine can cause environmental contamination and severe health hazards on human life. Different sensing technologies are used to detect and estimate hydrazine concentration in the atmosphere such as soil and water etc. Among these techniques, electrochemical technology shows high sensitivity and selectivity towards the detection of hydrazine.

Keywords

Hydrazine, Electrochemical, Colorimetric, Amperometry, Electrocatalytic Activity, Nanomaterial, Surface Plasmon Resonance

Contents

Materials Research Forum LLC
https://doi.org/10.21741/9781644901175-5

1.   Introduction

Hydrazine (N_2H_4) is the combination of two ammonia (NH_3) molecules bonded nitrogen-nitrogen single bond (N-N) with the removal of hydrogen (H_2). In hydrazine single N-N bond length is 1.45 A^0 and nitrogen-hydrogen (N-H) bond length is 1.05 A^0. N_2H_4 has sp^3 hybridization, 3 sigma (^{6}) bonds, and tetrahedral structure. Its IUPAC name is dinitrogen tetrahydride and simple form is known as diamine or hydrazine (N_2H_4) [1]. Hydrazine structure is shown in Fig. 1.

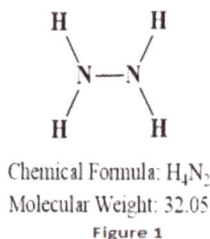

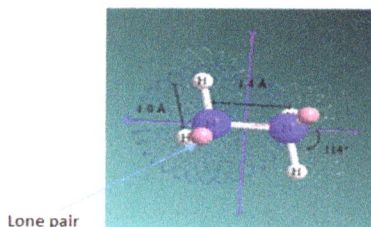

Chemical Formula: H_4N_2
Molecular Weight: 32.05
Figure 1

Lone pair

Fig. 1 Structure of hydrazine.

Hydrazine isa polar inorganic compound as nitrogen (N) is higher in electronegativity compared to hydrogen (H). So, the electron density is not equally distributed which gives hydrogen bonding. Hydrazine is soluble in water but not in organic compounds except the compounds containing hydrogen bonding. It is an alkaline liquid but weakly basic compared to ammonia. Hydrogen peroxide structure resembles N_2H_4 structures as shown in Fig. 2 [2, 3].

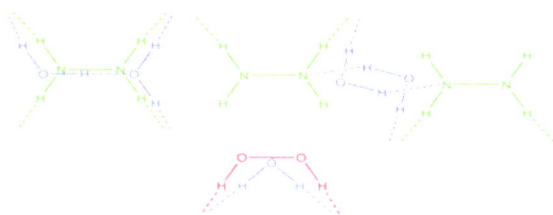

Fig. 2 Bonding structure of hydrazine.

1.1 Occurrence and preparation of hydrazine

The natural occurrence of hydrazine is very rare, and it is sporadic to extract the hydrazine from plant and minerals. Some exceptional natural source such as mushrooms, azotobacter, burley tobacco, flue-cured tobacco, snuff tobacco and bright tobacco are used to produce hydrazine. These natural sources contain a very small quantity of hydrazine [4,5]. Synthetic processes are used for the production of hydrazine, the following reaction is used for the production of N_2H_4 [6].

The reaction of chlorine with ammonia to form N_2H_4

$$2NH_3 + Cl_2 \longrightarrow N_2H_4 + 2HCl$$

Raschig synthesis reaction

$$NH_3 + NaOCl \longrightarrow NH_2Cl + NaOH$$

$$2NH_3 + NH_2Cl \longrightarrow N_2H_4 + NH_4Cl$$

The mixing of sodium hydroxide, 20% urea and chlorine to produce N_2H_4 [7].

$$(NH_2)_2COH + Cl_2 + 4NaOH \longrightarrow N_2H_4 + 2NaCl + Na_2CO_3 + 2H_2O$$

1.2 Physicochemical properties of hydrazine

Anhydrous hydrazine (N_2H_4) is a fuming-colorless, oily liquid having a density equal to water ($1g/cm^3$). Its melting point is 2°C and the boiling point is 114°C. It has an ammonia-like odor and gives a blue or violet flame in pure form. It is highly toxic and it mixes well with water resulting in $N_2H_4.H_2O$ hydrates and alcohols. It gives out salt which means that it behaves as a weak base like ammonia (NH_3). N_2H_4 and its derivative is the strongest reducing agent, it gives vigorous reaction with metal and oxidants [8].

1.3   Application of hydrazine in a different field

Hydrazine is an inorganic oily liquid compound that is used in many different industries and has attracted the attention of the scientific community. Hydrazine is used in many industries such as agriculture, power generation, pharmaceutical, aerospace, catalysis, and chemical industries. In the chemical industry N_2H_4 is used for different purposes like corrosive preventer, in cancer drugs used as a intermediate, antioxidant, foaming, catalysis, explosive, insecticide, herbicide and oxygen scavenger [9, 10].

1.4   Used in thermal energy

Hydrazine is unstable in pure form but stable in aqueous conditions. It is used in boiler feedwater for the removal of oxygen from the boiler water. Hydrazine is used to prevent the scaling of calcium, magnesium and phosphate salts in boiler pipes, antimicrobial activity, and corrosion. When N_2H_4 reacts with oxygen it decomposes into the H_2O and nitrogen [11]. The reaction is given below:

$$N_2H_4 + O_2 \longrightarrow N_2 + 2H_2O$$

1.5   Used as aircraft fuel

F-16s (C, D) Block 52 (Single seater and double seater) aircraft have a main following element (electronic device system, turbine drive component, transportable hydrazine tank, and nitrogen tank). Hydrazine is used as a fuel in a jet engine, N_2H_4 react with iron oxide and on decomposition gives gases and extreme thermal energy, and these gases rotate the turbine engine in 1-2 second at 75000 rpm. The gases produced from hydrazine decomposition generate heat energy having a temperature of 850 °C [6].

1.6   Health hazards

The wider use of hydrazine in many industries is directly or indirectly polluting the environment, living organisms and also aquatic life. Occupational safety and health administration (OSHA) define the permissible exposure limit (PEL) for 8 hours is 1 ppm. According to material safety data sheet (MSDS) and United State Environmental Protection Agency (USEPA), hydrazine poses several life threats such as genotoxic, hepatotoxic, neurotoxin, immunotoxin, carcinogenic, mutagenic, blood irregularity,

Toxic Gas Sensors and Biosensors
Materials Research Forum LLC
Materials Research Foundations **92** (2021) 139-156
https://doi.org/10.21741/9781644901175-5

several adverse effects to DNA, lungs, liver, kidney, central nervous system, irritation in eyes, nose, shakiness, vomiting, pulmonary edema, and unconsciousness. It is dangerous to aquatic life; when N_2H_4 reacts with oxygen to form H_2O and nitrogen.

$$N_2H_4 + O_2 \longrightarrow N_2 + 2H_2O$$

This reaction shows that all oxygen reacts with hydrazine and no more oxygen is left for aquatic life that is dangerous to all aquatic life. This causes the death to all aquatic life [12-14].

2. Different method used for the detection of hydrazine

In view of the above mentioned health hazard, it is necessary to monitor the quantity of hydrazine and protect the environment, human life, and aquatic life form. For the identification and estimation of hydrazine concentration in the atmosphere, water, and reduce threat to humans, it is apparent and obligatory to shape an effective, fast and extremely sensitive senor. In the literature review, some technique are reported for the qualitative/quantitative detection of hydrazine as shown in Fig. 3 [15, 16].

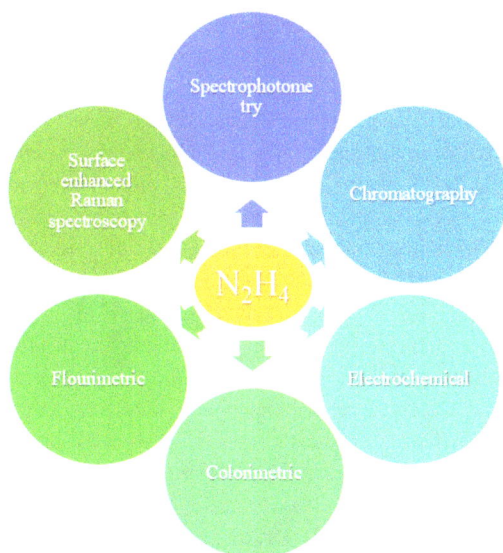

Fig. 3 Different techniques for sensing of hydrazine.

2.1 Electrochemical techniques

Electrochemical techniques detect analytes through the oxidation/reduction of analyte at the surface of the electrode. The interaction of analytes and the electroactive species results in the generation of electrochemical signals. The electrochemical signals are directly related to the concentration of the analytes. In electrochemical techniques potentiometry, voltammetry, and amperometry is used for the determination of analytes. [17]. These techniques have some advantages over conventional techniques Gas chromatography-Mass spectrometry(GC-MS), High performance liquid chromatography – Mass spectrometry (HPLC-MS), and spectrophotometer) for example low cost, high sensitivity, selectively, repeatability, reproducibility, low limit of detection, easy to handle. The demand for the development of electrochemical sensors has increased due to the upgradation of electronic circuit to allow some specific measurements of electro-analytical technique to be controlled by microprocessors or a computer [12]. These electro-analytical techniques are used in the sensing of different fields such as gas sensors, smoke sensors, biosensors, heavy metal sensing, sensing of an organic compound and electro-optical sensors, etc. [18]. In order to achieve better results of electrochemical sensors, it is possible to use nanomaterials that play a critical role in the detection of target analytes due to the larger surface area, numerous active sites, extraordinary catalytic properties. Redox reactions involve measuring the potential (potentiometry), current (amperometry) and conduction between electrodes (conductometry) and resistance (electrochemical impedance spectroscopy) in electrochemical sensors. The electrochemical sensor system consists of four main components; i- analyte ii- transducer iii- signal processing unit iv- display screen [19]. The electrodes are modified through different materials (carbon materials, graphene, reduced graphene oxide, conductive polymer, metal/metal oxide and metal alloy, nanocomposites of metal with carbon material and hybrid nanocomposites (metal, carbon material and conductive polymer), and these modified electrodes are used for the sensing of different analytes [20]. The nanocomposites show high sensitivity, selectivity, reproducibility and high limit of detection [21]. Different conventional techniques (GC, HPLC, and spectrophotometry) are used for the determination of hydrazine but electrochemical and fluorescence techniques are more sensitive for the detection of hydrazine [22, 23]. Different nanomaterials have been used for the preparation of the electrochemical sensor for hydrazine. The fabrication of the electrode with sensitivity, selectivity, and repeatability amperometric sensor based on rGO/Pt-TPP nanocomposite for the detection of hydrazine [24]. Detection of hydrazine in lake and tap water sample using the Au@Pd/CB-DHP hybrid nanocomposite by amperometric techniques. The fabrication of a modified electrode with high sensitivity and wide linear range (0.01-

150mM) based on Au-Cu NPs supported on P nano zeolite for the detection of N_2H_4 using amperometry [25]. The conductive substrates with deposition of nanomaterials have attracted attention toward the sensing of hydrazine. The modified electrode coated with three-dimensional interconnected $Co(OH)_2$ nanosheet is electrodeposited on Ti mesh surface for the detection of hydrazine by using amperometric techniques. The schematic diagram is shown in Fig. 4 [26].

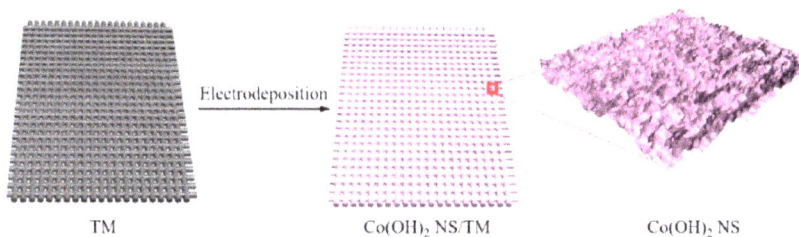

TM $Co(OH)_2$ NS/TM $Co(OH)_2$ NS

Fig. 4 The fabrication of $Co(OH)_2$ NS/TM [26].

The as-prepared nanosheets show the linear range from 5 μM to 3.0 mM and Limit of detection (Low limit of detection (LOD) of 1.98 μM. Electrode modified with nanocomposite (PtNPs-rGO-MWCNTs) synthesized with chemical reduction method shows good stability and high sensitivity for the amperometric detection of hydrazine [27]. The formation of highly electroactive nanocomposite material PQQ-modifiedmulti-walled carbon nanotube was applied for the detection of hydrazine by chronoamperometry as shown in Fig. 5 [28].

The prepared Pd/β-SiCNW-NC mesoporous composite was used for the amperometric determination of N_2H_4. The hybrid hetero-structure $Co(OH)_2$-Ni has low oxidation potential for the detection of hydrazine by linear sweep voltammetry (LSV) [29]. Fabrication of the Nd-Gd-TNT/Pd modified Au electrode shows good reproducible selectivity for the determination of hydrazine using cyclic voltammetry [30]. For the electrochemical sensing of N_2H_4 by cyclic voltammetry, the electrode was modified by hybrid AuNPs-NiO nanosheets with excellent sensitivity [31]. Hydrazine was determined by CDs-Cu_2O/CuO nanocomposite modified electrode using the cyclic voltammetry technique. It gave a good low limit of detection and high sensitivity for hydrazine detection [32].

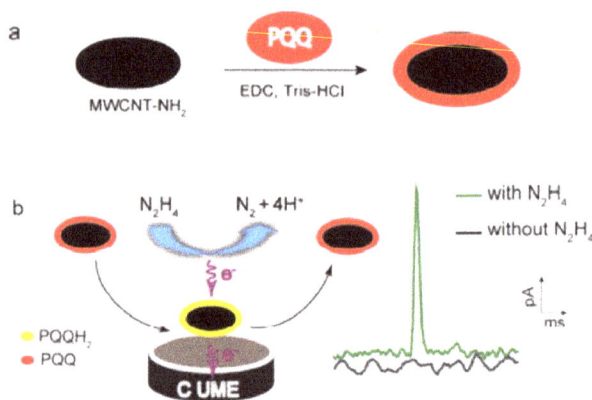

Fig. 5 The schematic diagram of a) The synthesis route of PQQ-modified MWCNT-NH₂(b) Electrocatalytic responses of hydrazine at carbon fiber ultra-micro electrode and current responses [28].

2.2   Fluorimetric techniques

Fluorescent sensors have excellent sensing applications. It usually consists of an element of recognition to detect target analytes as well as a fluorescent transducer (typically a fluorescent pigment). When analyte interacts with the recognition feature, it causes shifts in the wavelength or strength of the fluorescent signals. Various forms of synthetic fluorophores have been used. Different material have been used for the synthesis of novel fluorescent probe, some example are given here [33]. Fabrication of novel naphthalimide-based highly sensitive fluorescent probe for the detection of hydrazine using fluorimetry techniques [34]. With none as accepting group, imidazo [1, 2-*a*] pyridine as dye, a new sensor for hydrazine was investigated. The carbonyl group, which serves as a fluorescent quenching group in the sensor, can react with hydrazine to form a pyrazole, thus allowing fluorescent turn-on type signaling. With this sensor, selective detection of hydrazine could be realized in the presence of common small molecules, metal ions and anions in aqueous environment [35]. Detection of hydrazine through fluorimetric techniques using hybrid material Imidazo [1,2-α] pyridine [20]. A new colorimetric and near-infrared (NIR) fluorescent probe CF-1 based on a seminaphthorhodafluor dye was successfully designed and used for hydrazine determination. Upon reaction with N_2H_4, probe CF-1 showed obvious *off-on* NIR emission spectrum centered at 657 nm, as well as a distinct color change that can be distinguished by the naked eye. The results of fluorescence spectrum experiments indicated that probe CF-1 has high selectivity and low detection limitation (40.6 nM in the solution) [35].

2.3   Colorimetric sensing of hydrazine

The future of the sensor is focused on simplicity, cost effectiveness, and high sensitivity response. Sensor technology for the fast detection of the analyte with high sensitivity and selectivity up to date is challenging. Colorimetric sensors have been widely accepted for high sensitive and selective response toward various analytes. A colorimetric sensor is used for the instantaneous detection of an analyte. Colorimetry detects a color change that results from a chemical reaction. It is widely used in sensing technology. The surface plasmon resonance phenomenon explains when the incident light strikes the surface and the collective oscillation of free electron occurs in the visible region [36]. The surface of the substrate is modified or functionalized with a functional group or nanoparticle is based on the molecular interaction [37].

There are many challenges for the development of colorimetric sensors with characteristics like high sensitivity, selectivity, accuracy, precision, and linearity. The sensitivity is explained as the detection of the analyte at a very low concentration. Nanotechnology has a role in the development of the colorimetric sensor for the detection of hydrazine. For that several nanocomposites have been reported and nanoparticles showed high sensitivity and selectivity [38]. Nanoparticles have unique chemical and physical properties and are a promising material for detection and remediation of environmental pollutants [39]. Among these gold nanoparticles (AuNPs) are reported to show unique optical property which is known as localized surface plasmon resonance (LSPR). When light metal interaction occurs, it is known as plasmonics. The small size of gold nanoparticles shows a distinctive color (red or blue) and exhibits a strong UV-vis extinction band. The surface plasmon resonance band of AuNPs is located in the visible region and depends on the shape, size, crystallinity and interparticle space [40].

Behrooz Zargar and Amir Hatamie [41] have reported the gold nanoparticles for the detection of hydrazine in water samples. Fig. 6 [42] shows the schematic mechanism of colorimetric detection of hydrazine in which reduction of $AuCl^-$ to gold nanoparticles by hydrazine compound produced surface plasmon resonance peak. The redox reactions occur in the sample as aresult gold nanoparticles formed that were identified by measuring the localized surface plasmon resonance absorption. When the concentration of hydrazine is increased the LSPR show the linear response. The linear response range is 6.0 to 40.0 x10^{-6}M with the limit of detection 1.1x10^{-6}M

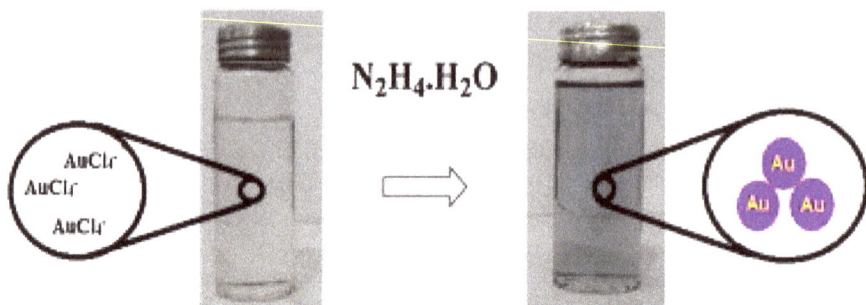

Fig.6 Colorimetric detection of hydrazine hydrated [41].

Yi He et al. [42] has reported the novel and low cost nanocomposite for the detection of hydrazine with high selectivity and sensitivity using MnO_2 nano-sheet as a probe. In the absence of hydrazine MnO_2 nano-sheet exhibit an ultraviolet-visible light at 370 nm absorption peak due to d-d transition and when hydrazine is added to MnO_2 sheet is converted into $Mn(OH)_2$ nanoparticles then absorption at 370 nm decreases. At this timeit is observed that yellow color is changing into colorless. The observed process is shown in Fig. 7.

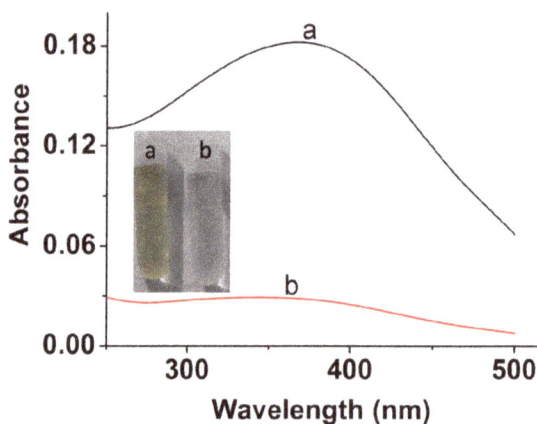

Fig. 7 UV-vis spectra of MnO_2 nanosheet solution in the absence (a) presence of hydrazine (b). The inset picture shows the sample[42].

The change in the spectrum shows the detection of hydrazine. The nanocomposite shows the linear range from 0.5 to 20 µM and the limit of detection of 0.4 µM for hydrazine. The small amount of concentration wasobservedby the naked eye in the limit of 11 µM.

2.4   Surface enhanced-Raman spectroscopy for hydrazine detection

Sensitive detection of the target analyte is important for sample analysis and mechanism study among the myriad of other components. Raman scattering is the phenomenon in which the inelastic scattering of photon reflects the vibrational modes of the interacting molecules [43]. Raman spectroscopy has many features like frequency shift, provides fingerprints and information of analyte chemical structure. Fleischmann et al. [44] found that the roughened silver electrode adsorbed the Raman signal and significantly enhanced. This phenomenon was later known as Surface Enhanced Resonance Spectroscopy (SERS), this described the unusual Raman signal enhancement when molecules are adsorbed on the metal surface of nanostructure. SERS is spectroscopic techniques for the detection of hydrazine. There is some issue of low affinity with gold and silver suitable for SERS detection. This can be solved by reacting with derivative reagents that can be strongly bound to the SERS substrate. The gold or silver strongly binds with the thio or amino group for the SERS substrate. This method is easy and highly selective and sensitive[45]. Camden et al. [46] have reported the detection of hydrazine in which hydrazine reacted with ortho-phthaldialdehyde to form phthalazine. It has better results toward the SERS substrate than hydrazines have shown is Fig. 8. The limit of detection was reported as 8.5 x 10^{-11} M. This technique is affordable and commercially available and the whole process is completed in 15 minutes.

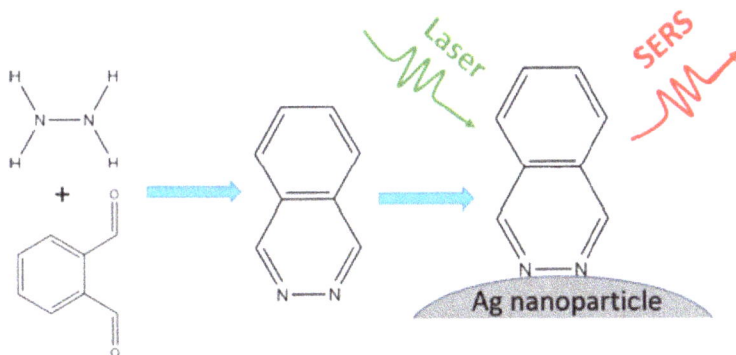

Fig.8 Schematic diagrams of SERS based hydrazine-detection [46].

2.5   Chromatography technique for the sensing of hydrazine

Chromatography is an analytical technique used for separating a mixture of chemical substance into its individual components and the components are analyzed thoroughly. Chromatography has many types like liquid chromatography, gas chromatography and ion-exchange chromatography; all have the same basic principle. These techniques are used for the sensing of hydrazine. Higher degrees of concentration in the preliminary stages of the sample is detected by chromatography techniques. The combination of selectivity and sensitivity makes the chromatography preferred for the sensing of hydrazine [47]. Sami Selmi et al. [48] have reported the chromatographic technique for the sensing of hydrazine in the water. Gas chromatography is used in that process. N_2H_4 is converted into azine that is determined by the gas chromatography using nitrogengas as a carrierand detected on the detector. The limit of detection is 0.1 ppb (part per billion).

2.6   Spectrophotometry sensing of Hydrazine

Hydrazine and its derivative are detected by spectrophotometry. It was the first analytical technique for the sensing of hydrazine compound at the level of 10^{-5} M. This method has many advantages such as low-cost equipment, a large number of users and simple analysis, there is no need fora skilled person. The hydrazine does not adsorb light in UV and near visible region while for the spectrophotometric measurement occur in this region. It is mandatory to carry out a chemical reaction that gives absorbing products.The spectrophotometric reaction determines the sensitivity and selectivity of the analysis [49]. The spectrophotometric sensor is used to detect the hydrazine in air. Xu Yao et al. [50] have reported the solution adsorption/spectrophotometry method for the determination of hydrazine in air. This method was fast, highly accurate and relatively simple in operation

Conclusions

Hydrazine is an inorganic oily liquid compound that is used in many different industries. Hydrazine causes environmental contamination and has severe effects on human life. For the identification and estimation of hydrazine concentration in the atmosphere, water, it is apparent and obligatory to shape an effective, fast and extremely sensitive senor. The monitoring of N_2H_4 in food, water and air is detected by different techniques. The different sensing techniques are discussed i.e. electrochemical sensor, colorimetric sensing, chromatography, fluorimetric and surface enhanced Raman spectroscopy. The nanocomposites material shows high sensitivity and selectivity towards the sensing of hydrazine.

References

[1] K. Patil, R. Mimani, Inorganic Hydrazine derivatives, synthesis, properties and applications: monograph. India: JohnWiley & Sons, Ltd Noida, 286 (2014).

[2] M. Yuan, D.B. Mitzi, Solvent properties of hydrazine in the preparation of metal chalcogenide bulk materials and films, Dalton Trans., (2009) 6078-6088. https://doi.org/10.1039/B900617F

[3] M.A. Navasardyan, L.G. Kuz'mina, A.V. Churakov, Unusual isomorphism in crystals of organic solvates with hydrazine and water, CrystEngComm, 21 (2019) 5693-5698. https://doi.org/10.1039/C9CE00978G

[4] G. Le Goff, J. Ouazzani, Natural hydrazine-containing compounds: Biosynthesis, isolation, biological activities and synthesis, Bioorg. Med. Chem., 22 (2014) 6529-6544. https://doi.org/10.1016/j.bmc.2014.10.011

[5] K. McAdam, H. Kimpton, S. Essen, P. Davis, C. Vas, C. Wright, A. Porter, B. Rodu, Analysis of hydrazine in smokeless tobacco products by gas chromatography–mass spectrometry, Chem.Cent. J., 9 (2015) 13. https://doi.org/10.1186/s13065-015-0089-0

[6] E. Janeba-Bartoszewicz, A. Rojewski, Analysis of hazards occurring during the use of hydrazine, J. Konse. Power. Trans., 25 (2018). https://doi.org/10.5604/01.3001.0012.4787

[7] N.H. Sabit, S.N.A. Jabbar, B.N. Basheer, N.M. Radhi, R.A.E. Abbas, S.T. Hawa, E.A. Abdullah, Preparation of hydrazine hydrate from urea and sodium hypochlorite, J. Iraqi Indust.Res. 4 (2017).

[8] O.V. Dorofeeva, O.N. Ryzhova, T.A. Suchkova, Enthalpies of formation of hydrazine and its derivatives, J. Phys. Chem. A, 121 (2017) 5361-5370. https://doi.org/10.1021/acs.jpca.7b04914

[9] R. Wahab, N. Ahmad, M. Alam, J. Ahmad, Nanorods of ZnO: An effective hydrazine sensor and their chemical properties, Vacuum, 165 (2019) 290-296. https://doi.org/101016./j.vacuum.2019.04.036

[10] N. Meher, S. Panda, S. Kumar, P.K. Iyer, Aldehyde group driven aggregation-induced enhanced emission in naphthalimides and its application for ultradetection of hydrazine on multiple platforms, Chem. Sci., 9 (2018) 3978-3985. https://doi.org/10.1039/C8SC00643A

[11] B. Bavarian, L. Reiner, J. Holden, B. Miksic, Amine base vapor phase corrosion inhibitor alternatives to hydrazine for steam generating systems and power plants, in: NACE 2018 Conference, April, 2018.

[12] S. Ghasemi, S.R. Hosseini, F. Hasanpoor, S. Nabipour, Amperometric hydrazine sensor based on the use of Pt-Pd nanoparticles placed on reduced graphene oxide nanosheets, Microchim. Acta, 186 (2019) 601. https://doi.org/10.1007/s00604-019-3704-2

[13] T. Beduk, E. Bihar, S.G. Surya, A.N. Castillo, S. Inal, K.N. Salama, A paper-based inkjet-printed PEDOT: PSS/ZnO sol-gel hydrazine sensor, Sens. Actuators B Chem., 306 (2020) 127539. https://doi.org/10.1016/j.snb.2019.127539

[14] H. Liu, H. Wang, G. Liu, S. Pu, H. Zhang, Ultrasensitive sensing of hydrazine vapor at sub-ppm level with pyrimidine-substituted perylene diimide film device, Tetrahedron, 75 (2019) 1988-1996. https://doi.org/10.1016/j.tet.2019.02.023

[15] F. Luan, S. Zhang, D. Chen, K. Zheng, X. Zhuang, CoS_2-decorated ionic liquid-functionalized graphene as a novel hydrazine electrochemical sensor, Talanta, 182 (2018) 529-535. https://doi.org/10.1016/j.talanta.2018.02.031

[16] S. Karakaya, Development of an amperometric hydrazine sensor at a disposable poly (alizarin red S) modified pencil graphite electrode, Monatsh. Chem., 150 (2019) 1911-1920. https://doi.org/10.1007/s00706-019-02513-4

[17] D. Martín-Yerga, Electrochemical detection and characterization of nanoparticles with printed devices, Biosens., 9 (2019) 47. https://doi.org/10.3390/bios9020047

[18] Q. Zhou, A. Umar, A. Amine, L. Xu, Y. Gui, A.A. Ibrahim, R. Kumar, S. Baskoutas, Fabrication and characterization of highly sensitive and selective sensors based on porous NiO nanodisks, Sens. Actuators B Chem., 259 (2018) 604-615. https://doi.org/10.1016/j.snb.2017.12.050

[19] B. Bansod, T. Kumar, R. Thakur, S. Rana, I. Singh, A review on various electrochemical techniques for heavy metal ions detection with different sensing platforms, Biosens. Bioelectron., 94 (2017) 443-455. https://doi.org/10.1016/j.bios.2017.03.031

[20] M.K. Rofouei, H. Khoshsafar, H. Bagheri, R.J. Kalbasi, Synthesis and characterisation of Ag-nanoparticles immobilised on ordered mesoporous carbon as an efficient sensing platform: application to electrocatalytic determination of hydrazine, Int. J. Environ. Anal. Chem., 98 (2018) 156-170. https://doi.org/10.1080/03067319.2018.1438419

[21] N. Teymoori, J.B. Raoof, M.A. Khalilzadeh, R. Ojani, An electrochemical sensor based on CuO nanoparticle for simultaneous determination of hydrazine and bisphenol

A, J. Iran. Chem. Soc., 15 (2018) 2271-2279. https://doi.org/10.1007/s13738-018-1416-x

[22] B. Luo, T. Wu, L. Zhang, F. Diao, Y. Zhang, L. Ci, J. Ulstrup, J. Zhang, P. Si, Monometallic nanoporous nickel with high catalytic performance towards hydrazine electro-conversion and its DFT calculations, Electrochim. Acta, 317 (2019) 449-458. https://doi.org/10.1016/j.electacta.2019.05.123

[23] C. Liang, H. Lin, Q. Wang, E. Shi, S. Zhou, F. Zhang, F. Qu, G. Zhu, A redox-active covalent organic framework for the efficient detection and removal of hydrazine, J. hazard. mater., 381 (2020) 120983. https://doi.org/10.1016/j.jhazmat.2019.120983

[24] S. Sakthinathan, S. Kubendhiran, S.M. Chen, M. Govindasamy, F.M. Al-Hemaid, M. Ajmal Ali, P. Tamizhdurai, S. Sivasanker, Metallated porphyrin noncovalent interaction with reduced graphene oxide-modified electrode for amperometric detection of environmental pollutant hydrazine, Appl. Organomet. Chem., 31 (2017) e3703. https://doi.org/10.1002/aoc.3703

[25] P.B. Deroco, I.G. Melo, L.S. Silva, K.I. Eguiluz, G.R. Salazar-Banda, O. Fatibello-Filho, Carbon black supported Au–Pd core-shell nanoparticles within a dihexadecylphosphate film for the development of hydrazine electrochemical sensor, Sens.Actuators B Chem., 256 (2018) 535-542. https://doi.org/10.1016/j.snb.2017.10.107

[26] J. Wang, T. Xie, Q. Deng, Y. Wang, Q. Zhu, S. Liu, Three-dimensional interconnected Co(OH)$_2$ nanosheets on Ti mesh as a highly sensitive electrochemical sensor for hydrazine detection, New J.Chem., 43 (2019) 3218-3225. https://doi.org/10.1039/C8NJ06008H

[27] H. Yu, S.-S. Wang, K.-L. Song, R. Li, A sensitive amperometric sensor for hydrazine based on Pt nanoparticles-reduced graphene oxide–multi-walled carbon nanotubes composite, Int. J. Env. Anal. Chem., 99 (2019) 854-867. https://doi.org/10.1080/03067319.2019.1616707

[28] F.T. Patrice, K. Qiu, L.-J. Zhao, E. Kouadio Fodjo, D.-W. Li, Y.-T. Long, Individual modified carbon nanotube collision for electrocatalytic oxidation of hydrazine in aqueous solution, ACS Appl. Nano. Mater., 1 (2018) 2069-2075. https://doi.org/10.1021/acsanm.8b00018

[29] E. Habibi, Mesoporous Pd| β-SiCNW-nC based home made screen printed electrode for high sensitive detection of hydrazine, Microchem.J., 149 (2019) 104004. https://doi.org/10.1016/j.microc.2019.104004

[30] H. Jiang, Z. Wang, P. Kannan, H. Wang, R. Wang, P. Subramanian, S. Ji, Grain boundaries of $Co(OH)_2$-Ni-Cu nanosheets on the cotton fabric substrate for stable and efficient electro-oxidation of hydrazine, Int. J. Hydrog Energy, 44 (2019) 24591-24603. https://doi.org/10.1016/j.ijhydene.2019.07.164

[31] K.M. Emran, S.M. Ali, H.E. Alanazi, Novel hydrazine sensors based on Pd electrodeposited on highly dispersed lanthanide-doped TiO_2 nanotubes, J.Electroanal.Chem., 856 (2020) 113661. https://doi.org/10.1016/j.jelechem.2019.113661

[32] R. Ahmad, T. Bedük, S.M. Majhi, K.N. Salama, One-step synthesis and decoration of nickel oxide nanosheets with gold nanoparticles by reduction method for hydrazine sensing application, Sens. Actuators B Chem., 286 (2019) 139-147. https://doi.org/10.1016/j.snb.2019.01.132

[33] G. Wei, L. Wang, L. Huo, Y. Zhang, Economical, green and rapid synthesis of CDs-Cu_2O/CuO nanotube from the biomass waste reed as sensitive sensing platform for the electrochemical detection of hydrazine, Talanta, (2019) 120431. https://doi.org/10.1016/j.talanta.2019.120431

[34] Y. Hao, Y. Zhang, K. Ruan, W. Chen, B. Zhou, X. Tan, Y. Wang, L. Zhao, G. Zhang, P. Qu, A naphthalimide-based chemodosimetric probe for ratiometric detection of hydrazine, Sens. Actuators B Chem., 244 (2017) 417-424. https://doi.org/10.1016/j.snb.2016.12.145

[35] X. Jia, X. Li, X. Geng, C. Nie, P. Zhang, C. Wei, X. Li, A seminaphthorhodafluor-based near-infrared fluorescent probe for hydrazine and its bioimaging in living systems, Spectrochim Acta A 223 (2019) 117307. https://doi.org/10.1016/j.saa.2019.117307

[36] J. Homola, I. Koudela, S.S. Yee, Surface plasmon resonance sensors based on diffraction gratings and prism couplers: sensitivity comparison, Sens. Actuators BChem., 54 (1999) 16-24. https://doi.org/10.1016/S0925-4005(98)00322-0

[37] A.P. VS, P. Joseph, K.D. SCG, S. Lakshmanan, T. Kinoshita, S. Muthusamy, Colorimetric sensors for rapid detection of various analytes, Mater. Sci. Eng.: C, 78 (2017) 1231-1245. https://doi.org/10.1016/j.msec.2017.05.018

[38] G.M. Whitesides, The origins and the future of microfluidics, Nature, 442 (2006) 368-373. https:///doi.org/10.1038/nature05058

[39] Y. Liu, G. Su, B. Zhang, G. Jiang, B. Yan, Nanoparticle-based strategies for detection and remediation of environmental pollutants, Anlst, 136 (2011) 872-877. https://doi.org/10.1039/C0AN00905A

[40] S.K. Ghosh, T. Pal, Interparticle coupling effect on the surface plasmon resonance of gold nanoparticles: from theory to applications, Chem. Rev., 107 (2007) 4797-4862. https://doi.org/10.1021/cr0680282

[41] B. Zargar, A. Hatamie, A simple and fast colorimetric method for detection of hydrazine in water samples based on formation of gold nanoparticles as a colorimetric probe, Sens. Actuators B Chem. 182 (2013) 706-710. https://doi.org/10.1016/j.snb.2013.03.036

[42] Y. He, W. Huang, Y. Liang, H. Yu, A low-cost and label-free assay for hydrazine using MnO_2 nanosheets as colorimetric probes, Sens. Actuators B Chem. B: Chem., 220 (2015) 927-931. https://doi.org/10.1016/j.snb.2015.06.025

[43] S. Schlücker, Surface-enhanced raman spectroscopy: Concepts and chemical applications, Angew. Chem. Int. Ed., 53 (2014) 4756-4795. https://doi.org/10.1002/anie.201205748

[44] M. Fleischmann, P. Hendra, A. McQuillan, Raman spectra of pridine adsobed at a silver electrode, Chem. phy. lett., 26 (1974).

[45] F. Wang, S. Cao, R. Yan, Z. Wang, D. Wang, H. Yang, Selectivity/specificity improvement strategies in surface-enhanced Raman spectroscopy analysis, Sens., 17 (2017) 2689. https://doi.org/10.3390/s17112689

[46] X. Gu, J.P. Camden, Surface-enhanced raman spectroscopy-based approach for ultrasensitive and selective detection of hydrazine, Anal. chem., 87 (2015) 6460-6464. https://doi.org/10.1021/acs.analchem.5b01566

[47] A. Smolenkov, Chromatographic methods of determining hydrazine and its polar derivatives, Rev. J. Chem., 2 (2012) 329-354. https://doi.org/10.1134/S2079978012040048

[48] S. Selim, C.R. Warner, Residue determination of hydrazine in water by derivatization and gas chromatography, J. Chromatogr. A, 166 (1978) 507-511. https://doi.org/10.1016/S0021-9673(00)95634-6

[49] A. Smolenkov, I. Rodin, O. Shpigun, Spectrophotometric and fluorometric methods for the determination of hydrazine and its methylated analogues, J. anal. chem, 67 (2012) 98-113. https://doi.org/10.1134/S1061934812020116

Toxic Gas Sensors and Biosensors | Materials Research Forum LLC
Materials Research Foundations **92** (2021) 139-156 | https://doi.org/10.21741/9781644901175-5

[50] X. Yao, G. Zhang, X. Wang, L. Wang, C. Fan, Solution absorption/spectrophotometry for determination of hydrazine in air, in: IOP conference series: Earth and environmental science, IOP Publishing, 2018, pp. 012036. https//doi.org/10.1088/1755-1315/113/1/012036

Toxic Gas Sensors and Biosensors
Materials Research Foundations **92** (2021) 157-196

Materials Research Forum LLC
https://doi.org/10.21741/9781644901175-6

Chapter 6

Graphene Nanostructures as Nonenzymatic Glucose Sensor

M. Saha*, J. Debbarma

Department of Chemistry, National Institute of Technology, Agartala-799046, Tripura, India

*mitalichem71@gmail.com

Abstract

The increasing demand for the development of highly selective and sensitive nonenzymatic electrochemical sensors for the qualitative and quantitative analysis of glucose in pharmaceutical, clinical and industrial sectors has gained enormous attention towards the use of graphene and its derivatives. This chapter describes the efficient development of electrochemically active nonenzymatic glucose sensors using graphene and its composites, achieving high sensitivity, stability, low detection limit, wide linear range and reproducibility.

Keywords

Nonenzymatic Sensor, Electrochemical Detection, Glucose, Graphene, Nanocomposites

Contents

Materials Research Forum LLC
https://doi.org/10.21741/9781644901175-6

1. Introduction

Carbon based nanostructures specially graphene and its derivatives are found to be extremely efficient for the analysis of biomolecules as it enhances the properties when combined with other metal or metal oxide based nanoparticles. Among all, graphene and its derivatives are considered to be a useful guide in the fabrication and development of next generation nonenzymatic glucose sensors. Diabetes is a condition marked by the inability of the body to manage the level of glucose in the blood and it is one of the leading causes of death and disability. The regular monitoring of blood is essential to know the glucose concentration which can greatly reduce the risk of diabetes mellitus, damage of tissues, stroke, blindness, kidney failure and heart disease. Thus, in the field of biosensor, the top issue is to develop a glucose sensor that is ideal in nature. Various processes such as electrochemical, conductometric, optical, fluorescent and colorimetric methods have been developed but in contrast to other available methods, electrochemical methods have attracted much more attention due to their unbeaten sensitivity, selectivity, lower detection limits, long term stability, faster response times and inexpensiveness. Hence electrochemical sensors have been widely used in clinical diagnosis, food detection, medical analysis and many more.

2. Electrochemical determination of glucose

The electrochemical sensors are said to be standard sensors for the application of glucose determination where the sensors show quite low detection limit with enhanced response times and these are cheaper compared to other sensor based available detection mechanisms. In electrochemical detection process, two specific methods have been used namely potentiometry and amperometry. In potentiometric sensor, a zero current is applied to measure the potential difference between the working electrode and reference electrode. It has been shown that the potential value of working electrode changes due to changes of glucose concentration and the sensor can also measure 10 µM or higher concentration of glucose. On the other hand, a constant bias potential is applied in amperometric glucose sensors and the resulting current is measured which scales linearly with glucose concentration. It also possesses almost similar glucose detection limit in the µM to M range as the potentiometric counterparts. Generally it can be classified into two categories enzymatic and nonenzymatic glucose sensors. Enzymatic biosensors are fabricated by immobilizing the enzymes on the surface of electrodes which can detect the analytics through the reaction and interaction with the analytics during electrochemical process. Enzyme based sensors show high sensitivity and good selectivity but their activity is affected by the environmental factors i.e. pH and temperature, leading to the limited lifetime of the sensors. To overcome these drawbacks, extensive attention has been paid on the nonenzymatic biosensors. The nonenzymatic glucose sensing depends on the electrochemistry of glucose i.e. oxidation as well as reduction which is a rapid and cost effective approach. The electrocatalytic performance of nonenzymatic sensors is slightly inferior to enzymatic sensor regarding the stability, detection range and repeatability. To fabricate nonenzymatic biosensors, numerous types of nanomaterials have been used till now including noble metals, quantum dots, magnetic nanoparticles, metal oxides, etc. Nanoparticles perform various roles such as catalyst towards electrochemical reaction, enhancement of electron transfer, immobilization of biomolecules and acting as reactant, whereas metal and metal oxide nanoparticles are highly stable and conductive.

3. Carbon nanostructure based non-enzymatic glucose sensors

Carbon is considered to be the most useful material in electrocatalysis and electroanalysis in nonenzymatic glucose sensing nowadays. Carbon based nanostructure especially carbon nanotubes are of great interest for the nonenzymatic electrode design owing to the unique electrical, electronic, mechanical and chemical properties. The nano-sized carbon nanotubes are found to be more convenient as nanoelectrodes, as their electrical properties can easily promote transfer of electrons during reactions. As carbon nanotubes can combine with metal nanoparticles for electrode design, it make them extremely attractive for new types of electric, optic and electrochemical biosensors because metal nanoparticle has extraordinary and physicochemical characteristics which enhances the surface area of carbon nanotubes as well as chemical, electrochemical stability, electrical conductivity with acceptable biocompatibility. In recent years considerable attention has been paid using graphene nanostructures and its composites for developing economical yet rapid glucose sensors. Being an unrolled carbon nanotubes, graphene has a high surface area which can serve as a superior platform for fabricating new composites in glucose sensing. Moreover the unique two dimensional crystal structure of graphene makes extremely attractive for metal and metal oxide nanoparticles as a supporting material due to its electrochmical and photochemical properties.

4. Graphene and its derivatives as potential non-enzymatic glucose sensor

Graphene based enzymatic glucose sensor was first reported in 2009 by Changsheng Shan and coworkers [1]. However, the use of graphene or its derivatives for non-enzymatic detection of glucose electrochemically, was reported by Hongcai Gao &

coworkers [2] in 2011, who obtained a high sensitive value of 20.42 μA cm^{-2} mM^{-1} using modified graphene and platinum-nickel nanoparticles with rapid response and excellent reproducibility as compared to other platinum based nanocomposites. In the same year, Jun-Yung Sun et al. [3] developed graphene/Ni(II)/quercetin composite which exhibited detection limit of 0.5 μM, S/R=3, good reproducibility and fast response towards the detection of glucose. In 2012, graphene/cuprous oxide nanoparticle based sensor was prepared by Yong Qian and coworkers [4] which showed higher sensitivity and excellent selectivity than chemically reduced graphene or Cu_2O in alkaline media as well as linear range and limit of detection (1.2 μM, S/N=3). The Xiao-Chen Dong & coworkers [5] reported 3D graphene/cobalt oxide electrode for the detection of glucose with ultrahigh sensitivity of 3.39 mA mM-1 cm^{-2} and lower detection limit of <25 nM (S/N = 8.5). Yu-Wei Hsu & coworkers [6] synthesized CuO/graphene composites which provided superior sensitive range of 1065.21 μA mmol-1 L cm^{-2} due tosmaller particle size of CuO (15.75 mm). The detection limit was observed as 1 μmol L^{-1} with the linear range from 1 μmol L^{-1} to 8 mmol L^{-1}. Fen-Ying Kong et al. [7] fabricated graphene oxide-thionine-gold nanostructure and the resultant sensor showed a lower detection limit of 0.05 μmol/L. Cai Yu Lu et al. [8] prepared graphene-gold nanoparticle-carbon nanotube nanohybrid, exhibiting linear behavior in the concentration range from 5 μM to 1 mM with a detection limit of 1.5 μM due to the large surface area of graphene/carbon nanotube. Liqiang Luo and coworkers [9] proposed modified CuO-graphene-glassy carbon electrode which exhibited fast response of 5s within the glucose concentration range from 2 μM to 4 mM, detection limit of 0.7 μM (S/N=3) as well as high sensitivity of 1360 μA mM^{-1} cm^{-2}. Jing Luo et al. [10] developed a novel Cu nanoparticles/graphene composite by simple one step pathway which displayed good catalytic activity, high sensitivity of 607 μA mM^{-1} and a low detection limit (200 nM). Microwave irradiation assisted process for the preparation of reduced graphene oxide/nanospheres of Ni based hybrids were developed by Zhigang Wang et al. [11]. Xuewan Wang et al. [12] synthesized graphene cobalt oxide based needle electrode which can directly detect glucose with lower detection limit less than 10 μM. In 2013, the graphene-CuO nanocomposites was fabricated by Yancai Li & coworkers [13], and this sensor gave high sensitivity of 1480 μA mM^{-1} cm^{-2}, fast response time of 3s, linear dependence in the range from 2.0 μM to 0.06 mM and low detection limit of 0.29 μM (S/N=3). Zonghua wang & coworkers [14] pasted polymer assisted copper nanoparticles on graphene surface and the electrode enhanced the sensitivity, selectivity, detection limit (0.08 μM) and linear range (0.3 μM-0.6 mM). Zengjie Fan & coworkers [15] synthesized a graphene transparent electrode with Cu nanowires, which displayed satisfactory sensitivity (1100 μA/mM cM^2), good stability and selectivity compared to indium tin oxide and glassy carbon electrode for glucose detection in blood and food. To investigate

Toxic Gas Sensors and Biosensors   Materials Research Forum LLC
Materials Research Foundations **92** (2021) 157-196   https://doi.org/10.21741/9781644901175-6

the electrocatalytic activity for glucose detection, M.F. Hussain and coworkers [16] fabricated modified graphene oxide by Pd nanoparticles to get excellent sensitive value of 512.5 μA mM^{-1} cm^{-2}, long term stability and working potential. As compared to other unsupported Cu_2O nanocubes, Minmin Lui and coworkers [17] proposed wrapped Cu_2O nanocubes/graphene which exhibited higher catalytic activity with sensitivity, detection limit of 3.3 μM (S/N=3) and range of linearity from 0.3-3.3 mM. Wenbo Lu et al. [18] prepared nickel based metal/ polymer nanoparticle/ reduced graphene oxide composite for achieving satisfactory sensitivity with detection limit from 0.01 to 8.75 mM. In the same year, Peng Si and coworkers [19] established Mn_3O_4/3D graphene sensor, showing good response towards the sensitivity value of 360 μA mM cm^{-2}. The intercalated graphene with ionic liquids and nickel hydroxide based sensor was proposed by Lu Wang & coworkers [20]. By using this electrode the current response with linearity was observed in a range from 0.5-500 μM and sensitivity of 647.8 μA mM^{-1} cm^{-2}. The modified electrode of CuO nanoneedle/carbon nanofibre/graphene composites were reported by Daixin Ye & coworkers [21] with useful linear range (1 μM-5.3 mM), rapid response (less than 2 s), low detection limit (0.1 μM). Baiqing Yuan and coworkers [22] prepared a Cu_2O/NiOx graphene based modified glassy carbon electrode and this electrode exhibited high catalytic activity due to synergistic effect of Cu_2O and NiOx and the observed value of sensitivity in case of spherical electrode (285 μAmM^{-1}cm^{-2}) was found to be higher as compared to octahedral electrode (23 μAmM^{-1}cm^{-2}). Graphene, Pd/Pt nanomaterials based biosensor was also reported by Hui Zhang et al. [23] via one-pot microwave heating for getting good results as compared to unsupported Pd, Pt nanomaterials. The range of linearity from 2 μM to 2.1 mM with less detection of 0.1 μM was observed using nickel nanoparticles with as reported by Yue Zhang & coworkers [24]. In 2014, Sushmee Badhulika and coworkers [25] fabricated multiwall carbon nanotube/graphene electrode decorated by platinum nanoflowers which responded linearly to glucose showing sensitivity of 11.06 μAmM^{-1}cm^{-2} at neutral pH. A graphene and Pt-Pd bimetallic nanocubes based modified glassy carbon electrode was prepared later on which showed good performance on comparing with other modified electrodes in glucose determination [26]. Yaolin Zhang & coworkers [27] prepared a nonenzymatic sensor using cobalt oxide nanoparticles/graphene oxide, getting superior electrochemical activity with detection maxima of 0.18 μM. Graphene assisted copper oxide nanoparticle electrodes were synthesized byAlizadeh and Mirzagholipur [28] which exhibited exceptionally large sensitivity of 2939.24 μA mM^{-1} cm^{-2}. Suqin Ci and coauthors [29] used cobalt oxide/graphene hybrids based electrode with rapid response (< 3 s), detection range (0.83 μM-8.61 mM), stability and good reproducibility.

Materials Research Forum LLC

https://doi.org/10.21741/9781644901175-6

To get protective and selective coating material, graphene derivative such as reduced graphene oxide has been widely used in biosensor. It was found that when reduced graphene oxide is mixed with a copper salt solution, the Cu^{2+} ion is selectively bonded with carboxylic group present in the reduced graphene oxide through a mutual interaction, as a result the electrocatalytic activity is increases. In 2014, reduced graphene oxide decorated by Pt-CuO nanoparticle composites was developed for glucose detection. This sensor showed very high sensitivity (3577 $\mu AmM^{-1}cm^{-2}$) with linear response upto 12 mM and detection of limit upto 0.01 μM (S/N=3) [30]. Xiaoxia Feng et al. [31] synthesized reduced graphene oxide decorated Cu_2O electrode by one-pot solvothermal method, the fabricated sensor showed good catalytic activity than pure Cu_2O nanowires towards direct oxidation of glucose i.e. high gradient of 80.17 $\mu A/mM$ with glucose concentration ranging from 10 μM-0.1 mM were obtained. A large value of sensitivity of 2042 $\mu AmM^{-1}cm^{-2}$ with linear range and detection limit were achieved by using electro reduced graphene oxide, multi walled carbon nanotube based nickel hydroxide sensor, which was developed by Wei Gao and coworkers [32]. Hu Yao-Juan and coworkers [33] developed flower shaped Pt-Au graphene based sensor for glucose detection. The results indicated that this decorated electrode showed good catalytic activity i.e. sensitivity (26.33 $\mu Acm^{-2}mM^{-1}$), low detection limit (4.0 μM) with wide linear range (1.0-25.0 mM) as well as stability. To compare the role of hollow nanostructure, Yaojuan Hu and coauthors [34] proposed graphene based hollow Pt-Ni nanostructure for glucose detection, which exhibited better linear range of 0.5-2.0 mM, sensitivity (30.3 $\mu AmM^{-1}cm^{-2}$) and good repeatability. Huanhuan Huo et al. [35] prepared Cu_2O/reduced graphene oxide composites in contrast to other reported Cu_2O based glucose sensor to get better results towards electrocatalytic detection of glucose with low detection limit of 0.5 μM. Nur Syakimah Ismail et al. [36] decorated the intercalation of graphene oxide nanoribbon with gold nanoparticle which achieved a linear range of 0.5 μm-10 mM, whereas copper nanoparticle assisted nitrogen doped graphene composites was achieved by Ding Jiang & coauthors [37], showing good selectivity,sensitivity of 48.13 μAmM^{-1}, low detection limit (1.3 μA, S/N=3) withvery fast response of less than 5s. Yong Kong & coworkers [38] presented nickel hexacyanoferrate/polyaniline graphene electrode with good catalytic performance including sensitivity (487.33 $\mu AmM^{-1}cm^{-2}$), linear range from 0.001-0.765 mM and detection limit (0.5 μM). Ionic liquid assisted Pt-Pd decorated on reduced graphene oxide composites displayed enhanced catalytic activity (linear range from 0.1-22 mM at 0V, considerable stability, response time within 3 s and good reproducibility) as compared with other Pt-Pd-graphene oxide composites [39]. Nanocomposite prepared by using reduced graphene oxide, copper coralloid granules and polyaniline responded towards the catalytic activity with sensitivity of 603.59 $\mu Acm^{-2}mM^{-1}$ with low detection limit of 1.34 μM (S/N=3), whereas reduced graphene oxide based nickel hydroxide composite exhibited sensitivity of 11.4 ± 0 mA mM^{-1} cm^{-2} and detection limit of 15 µM [40, 41]. Li Wang and coworkers [42] fabricated a composite with reduced graphene oxide, chitosan and copper-cobalt nanostructures and it displayed good sensitivity of 1921 µAcm^{-2}mM^{-1} and stad compare to other monometals such as Cu and Co. Glassy carbon electrode with copper oxide and reduced graphene oxide showed high sensitivity (2221 µAmM^{-1}cm^{-2}) and bigrange of linearity (0.4 µM – 12 mM), while graphene based nickel nanoparticle composites exhibited good sensitivity of 865 µAmM^{-1}cm^{-2}, limit of detection of 1.85 µM (S/N=3) 5-550 µM [43, 44]. The fabricated quantum sized SnO$_2$ and reduced graphene oxide composites, proposed by Yixing Ye and coauthors [45] showed good catalytic activity and the sensitivity was determined to be 1.93 A M^{-1} cm^{-2} with limit of detection as 13.35 µM. Beibei Zhan and coworkers [46] developed an electrode based on three dimensional graphene with nickel hydroxide and the electrode displayed low detection limit (0.34 µM), sensitivity (2.65 mAmM^{-1}cm^{-2}) and a wide linear range (1 µM – 1.17 mM). Compared to pure Cu$_2$O nanospheres, Dan-Ling Zhou and coauthors [47] constructed monodisperse porous Cu$_2$O nanospheres and reduced graphene oxide based electrode having fast response (within 3s), sensitivity (185 µAmM^{-1}), low detection limit (0.05 µM) and selectivity. In 2015, nickel oxide hollow spheres with reduced graphene oxide electrode showed very high sensitivity of 3796 µA mM^{-1}cm^{-2} and 2721 µAmM^{-1}cm^{-2} as proposed by Pan Lu and coworkers [48]. Aswathi R and coauthors [49] build up a glucose sensor using graphene/gold with exceptionally high detection limit of glucose of 10 nM as compared to the other electrode. The glucose sensor demonstrated by Junwei Ding & coworkers [50] using Cu$_2$O microspheres on graphene nanosheets had detection maxima of 7.288 x 10^{-4} mM with detection range from 0.001-0.419 Mm. In the same year, a composite based on graphene, Au nanoparticles, Ni-Al layered double hydroxides and single wall carbon nanotubes was established by Shuai Fu & coauthors [51] for nonenzymatic glucose detection with high sensitivity value of 1989 µA mM^{-1} cm^{-2} and low detection limit (1.0 µM). Ni Hui and coworkers [52] used reduced graphene oxide doped with poly (3,4-ethylenedioxythiophene) for achieving limit of detection of 0.8 µM (S/N=3) and good stability. Ni Hui and coauthors [53] introduced a composite based on graphene oxide/poly (3, 4 ethylenedioxythiophene)/Cu nanoparticle and it exhibited sensitivity up to 909.1 µA mM^{-1}cm^{-2}, low detection limit of 47 nm, fast response time (less than 1 s) and a wide linear range from 0.1µM-1.3 mM. It was further observed that three dimensional graphene-Cu hybrid sensor exhibited the detection limit of 18 nM (S/N =8.6) with ultra-high sensitivity of 7.88 mAmM^{-1}cm^{-2} [54]. Later on, porous reduced graphene oxide with cobalt and nickel was reported to possess excellent electrocatalytic activity towards glucose detection with low detection limit (0.078 µM) and sensitivities for two linear ranges (1753 µAmM^{-1}cm^{-2} and 954.7 µAmM^{-1}cm^{-2}), whereas polyvinylpyrrolidone-graphene nanosheets-chitosan-Ni nanoparticles based composite displayed good catalytic activity, achieving good sensitivity, detection limit (30 nM), wide linear range (0.1 μM to 0.5 mM), high stability, selectivity and reproducibility [55, 56]. Pooria Mn et al. [57] reported a nanocomposite using reduced graphene oxide-CuO-polypyrrole nanofibres and the sensor exhibited detection limit of 0.03 μM (S/N=3), stability, reproducibility and selectivity properties. Sreeramareddygari Muralikrishna et al. [58] designed a CuO nanobelts-graphene composite which showed higher sensitivity and selectivity as compared to other chemically reduced graphene oxide or CuO nanobelts in alkaline media. Reduced graphene oxide/ NiS prepared by Radhakrishnan and Kim [59] showed high sensitivity, selectivity and satisfactory reproducibility towards glucose detection. Three dimensional graphene/nickel hydroxide electrode based glucose sensor was decorated by Iman Shackary and coworkers [60] for achieving lower detection limit up to 24 nM, ultrahigh sensitivity upto 3.49 $mAmM^{-1}cm^{-2}$ due to the high surface area of fabricated electrode. Leila Shahriary and Athawale [61] built up an electrode by reduced graphene oxide/silver/silver oxide and recorded a detection limit of 0.060 mM (S/N=3) and sensitivity upto 32 $μAmM^{-1}cm^{-2}$. Honghui Shu and coworkers [62] synthesized Au-graphene composite in one step and it catalyzed the oxidation of glucose in neutral media, showing a linear range (0.1-16 mM), fast response time (about 3s) and the sensitivity was estimated as 4.56 $μAcm^{-2}mM^{-1}$. Farshad Tehrani and coauthors [63] developed a sensor decorated with graphene and copper nanoparticles and demonstrated high sensitivity upto 1101.3±56 $μA/mM.cm^{-2}$, limit of detection (0.025-0.9 mM), linear response from 0.1-0.6 mM and good stability. Recent studies demonstrated that doping of heteroatoms into graphene can modulate the electronic structure and properties and also enhance the interaction between the metal/metal oxide nanoparticles. The doping of hetero atom like sulfur in graphene and CuO was reported by Ye Tian et al. [64] and the sensor exhibited high sensitivity upto 1298.6 $μAmM^{-1}cm^{-2}$ and a rapid response time (2s) along with good selectivity, accuracy and stability. Another electrode coated with reduced graphene oxide and copper nanoparticles showed excellent catalytic activity i.e. high sensitivity up to 447.65 $μAmM^{-1}cm^{-2}$ with detection limit of 3.4 μM [65]. Li Wangand coworkers [66] synthesized electrode using reduced graphene oxide and Pt-nickel oxide nanoplate array in one step and it recorded a limit of detection as 2.67 (S/N=3), sensitivity as 832.95 $μAcm^{-2}mM^{-1}$ and range of the linearity was observed from 0.008 to 14.5 mM. The reduced graphene oxide and Cu_2O (size of 4 nm) based composite was prepared by Xiaoyan Yan and coauthors [67] for electrochemical glucose detection and the resultant elecrrode was found to be strongly sensitive due to the nano-sized Cu_2O. Suling Yang and coworkers [68] doped nitrogen in graphene and Mn_3O_4 nanoparticles and found that the electrode showed response current 10 fold higher than pure glassy carbon electrode towards glucose. In the same year, Suling Yang & coauthors [69]

Toxic Gas Sensors and Biosensors   Materials Research Forum LLC
Materials Research Foundations **92** (2021) 157-196   https://doi.org/10.21741/9781644901175-6

proposed nitrogen doped reduced graphene oxide and Mn_3O_4 nanoparticles composite which exhibited the response current of glucose 20 fold higher compared to pure Mn_3O_4 and the sensitivity was recorded as 0.026 µA/µM, low detection of 0.5 µM, S/N=3, fast response time with linearity of 1.0-329.5 µM. Xiaohui Zhang and coworkers [70] proposed modified electrode of graphene with copper-nickel oxide nanoparticles and the glucose sensor showed linear range of 16 mM (from 0.05-6.9 mM and 6.9-16 mM), sensitivity of 225.75 $µAmM^{-1}cm^{-2}$ and 32.44 $µAmM^{-1}cm^{-2}$. Copper oxide supported porous reduced graphene oxide electrode was developed by Yue Zhao and coauthors [71] which responded in glucose detection in a linear range from 0.001 to 6 mM and the limit of detection was found to be 0.50 µM. In 2016, Pranati Nayak and coworkers [72] proposed modified glassy carbon electrode/Pd nanoparticle/ graphene/wrapped CNT composite as sensing matrix, where the sensor possessed fast response time and good reproducibility, while R. Prasad & coauthors [73] used multi walled CNTs with reduced graphene oxide and nickel oxide to prepare the electrode. Seyed Morteza Naghib et al. [74] used Fe_3O_4 and gelatin and the developed electrode exhibited enhanced shelf life (> 2 month), low detection limit of 0.024 µM, wide linear range (0.1-10 mM) and high sensitivity. To overcome the drawbacks of high cost of rare metal precursors and low sensityvity, molecularly imprinted polymers were developed which may fulfill many requrements. Mohammad Masoudi Farid et al. [75] proposed a glucose sensor with polyvinyl acetate, CuO, graphene oxide nanoparticles and MnO_2 using molecular imprinting method and their investigation suggested that the electrode showed much higher response and linear range (0.5-4.4 mM) than the composite prepared by nonprinted method. An electrode based on reduced graphene oxide/cobalt oxide nanoparticle was prepared using electrodeposition methodand the electrochemical studies showed that the modified electrode also displayed low detection limit (1.4 µM, S/N=3), with good sensitivity value of 1.21 $µAmM^{-1}cm^{-2}$ [76]. Chung and Hur [77] used hydrothermal method to prepare graphene oxide hydrogels and Co_3O_4 nanoflowers based sensor which exhibited high sensitivity up to 492.8 $µAmM^{-1}cm^{-2}$, 20 times higher than pure graphene oxide hydrogels. Lele Ju and coworkers [78] prepared copper nanowires/reduced graphene oxide composites using wet chemical process in single srep. A low detection limit of 0.2 µM, high sensitivity of 1625 µA/(mM.cm^{-2}), linear range up to 11 mM and fast response (less than 2 s) towards glucose detection were observed with the sensor. Su-Juan Li and coauthors [79] synthesized a sensor using reduced graphene oxide, glassy carbon electrode, copper-cobalt oxide by electrochemical technique and the hybrid electrode showed better catalytic activity (linear range from 5-570 µM and detection limit is 0.5 µM) than other glassy carbon electrode modified with CuO or CoO. Yu Luo & coworkers [80] proposed a sensor which possessed good reproducibility and stability along with linear range from 0.00001-80 mM and detection limit is 0.0022 µM.

In the same year, another reduced graphene oxide composite decorated with spinel nickel-cobalt oxide nanowrinkles was established by Guangran Ma & coauthors [81]. The developed composite enhanced the catalytic activity compared to single component of NiO or CoO and a wide linear range were observed 2 μM, S/N=3 and 0.005-8.6 mM. Vikrant Sahu & coworkers [82] also proposed a reduced graphene oxide and CuO particles based composite, which displayed a good response with high sensitivity around 1500 μA/mMcm^2 and the response time was found as only 1s. Iman Shackery & coworkers [83] reported an economic method (chemical bath) to prepare porous graphene and copper hydroxide nanorods based electrode and it provided a high sensitivity of 3.36 mA mM^{-1} cm^{-2}, linear range from 1.2 μM – 6 mM, low detection limit of 1.2 μM, reproducibility and high selectivity. The same reserchers again reported the graphene and cobalt dihydroxide nanorods based composites using the method and this nanocomposite revealed a high sensitivity for glucose up to 3.69 mAmM^{-1}cm^{-2} with detection maxima of 16 nM [84]. Zohreh Shahnavaz and coauthors [85] modified reduced graphene oxide with zinc ferrospinels and glassy carbon, where the electrode displayed remarkable catalytic activity towards glucose with sensitivity of 1689.6 μA mM^{-1} cm^{-2} and the linear range from 1.0×10^{-2} to 12.5×10^{-2} M. Another modified reduced graphene oxide based composite was synthesized by Zongxu Shen and coworkers [86] and the electrode showed outstanding performance for glucose sensing with good linear range (1-710 μM), high sensitivity (1414.4 μAmM^{-1}cm^{-2}) and detection limit (0.37 μM). Nitrogen doped graphene with gold nanoparticle nanohybrid was later on synthesized by through seed assisted growth method and the resultant nanohybrids displayed detection limit of 12 μM, short response time (10 s) and linear range from 40 μM to 16.1 mM [87]. Chao Wang & coworkers [88] proposed silver-platinum anchored reduced graphene oxide composite by a galvanic replacement reaction and thermal reduction process. The results showed a fast response time (less than 3s), sensitivity up to 129.32 μAmM^{-1}cm^{-2} and low detection limit of 1.8 μM (S/N=3). Graphene oxide/copper nanoparticle/single wall carbon nanotubes composite, prepared by electrodeposition method provided large surface area, hence it showed sensitivity up to 930.07 μAmM^{-1}cm^{-2} with less detection of 0.34 μM (S/N=3) towards glucose sensor [89]. Siti Nur Akmar Mohd Yazid & coauthors [90] fabricated a sensor based on graphene/alkaline reduced cuprous oxide/modified glassy carbon which presented fast response time of 3s, high sensitivity upto 1330.05 μA mM-1 cm-2, detection limit of 0.36 μM and a wide linear range of 0.01 mM to 3.0 mM. Guisheng Zeng et al. [91] synthesized graphene nanosheets with nickel oxide nanoparticles composite through hydrothermal route. Compared with bare nickel oxide, the fabricated sensor could response to glucose much higher in a low detection limit of 5.0 μM, S/N=3 and a wide linear range from 5 μM to 4.2 mM. Cong Zhang and coworkers [92] introduced reduced graphene oxide/Au nanoparticles/manganese dioxide composite using ecofriendly pathway, while reduced graphene oxide/nickel/nickel nanoflower composite was introduced by Xiaohui Zhang & coworkers [93] for electrochemical glucose sensor and they proposed that the sensor can be utilized for glucose detection with sensitivity of 1997 µA/mMcm^{-2}, linearity range from 29.9 µM - 6.44 mM and a low value of detection of 1.8 µM, S/N=3. The detection of glucose using graphene derivatives increased continously and Anshun Zhao and coauthors [94] reported a flexible electrode constructed by graphene paper, nanoporous gold and scaffold with high loading of platinum-cobalt nanoparticle and its performance towards detection of glucose included sensitivity of 7.84 µAmM^{-1} cm^{-2}, low detection limit of 5 µM, S/N=3 and wide linear range from 35 µM-30 mM. Jianzhong Zhang and coauthors [95] fabricated reduced graphene oxide/cupric oxide particle modified electrode through fast microwave method which detected glucose in the concentration range of 0.50 µM to 3.75 mM and the detection limit was 0.10 µM, S/N=3. Weiren Zhang & coworkers [96] fabricated graphene/copper nanoparticles/polyaniline composite which showed sensitivity towards glucose oxidation within response time of 3s and limit of detection was found to be 0.27 µM, S/N=3. Xuming Zhuang et al. [97] proposed a sensor based on graphene oxide/nickel oxide nanoparticles/polyaniline nanowire/modified glassy carbon and the sensor showed higher catalytic activity than nickel oxide nanosheets sensor. In 2017, Lin Bao and coworkers [98] built up graphene frameworks with Co$_3$O$_4$ applying thermal explosion method and the glucose sensor gave lower detection limit (157 ×109 M, S/N=3) and sensitivity value of 122.16 µAmM^{-1}cm^{-2}. Ghanbari and Ahmadi [99] synthesized reduced graphene oxide/Au/polyvinyl nanofibre/NiO nanostructure based composite using electrochemical technique which displayed remarkable electrocatalytic performance showing low detection limit (0.23 µM, S/N=3), linear range from 0.09-6 mM, good selectivity and stability and fast response time compared to the spectrophotometric method. Al. Gopalan & coauthors [100] utilized hydrothermal method for preparing graphene hybrid nanosheets and bismuth oxychloride composite, where the composite showed good performance within wide concentration range from 500 µM to 10 mM, sensitivity (1.878 µAmM^{-1}cm^{-2} and 127.2 µAmM^{-1}cm^{-2}), reproducibility and good selectivity.

Tae-Hoon Ko and coworkers [101] prepared NiCo$_2$O$_4$/graphene nanohybrids and found that compared to pure graphene, this composite showed good catalytic activity towards glucose oxidation using concentration of 0-0.14 mM. Meixia Li and coauthors [102] reported nitrogen and sulfur codoped graphene and CuO nanoparticle based composite and the sensor displayed a considerable limit of low detection of 0.07 µM, high sensitivity of 1722 µA/mM/cm^{-2} as well as good stability, selectivity and reproducibility. Luba Shabnam and coworkers [103] developed a glucose sensor based on nitrogen doped graphene with dispersed copper nanoparticles by thermal annealing process at 900°C for 1 hour. The developed electrode presented a linear response of 0.01-100 µM with low detection threshold of 10 nM. Sreejesh M and coworkers [104] synthesized reduced graphene oxide and zinc oxide nanocomposites using microwave method. The nanocomposites enhanced the electrochemical properties significantly in glucose detection with lower limit of detection of 0.2 nM and large sensitivity of 39.78 mAcm^{-2} mM^{-1}. Zhimei Zhang & coauthors [105] built up few layer graphene/copper oxide composite which allowed glucose sensing with high sensitivity up to 3120 µAmM-1cm^{-2} and very fast response time of 2s. Sheeba Alexander and coauthors [106] proposed graphene oxide-molecular imprinted polymer based composite which possessed detection limit of 0.1 nM, while Paramasivam Balasubhramanian and coworkers [107] synthesized electrochemically reduced graphene oxide wrapped/copper nanoparticles composite and observed low detection limit of 49 nM, a wide linear range from 0.14 µM - 5091 mM as well as appreciable reproducibility. S. Darvishi & coworkers [108] prepared reduced graphene oxide/nickel nanoparticle/ethylene glycol composite for enzyme free electrochemical glucose detection. The composite showed higher catalytic performance in alkaline solution, where detection limit was 0.01 µM compared to reduced graphene oxide. Sathiyanathan Felix & coworkers [109] reported nitrogen doped graphene/ CuO nanocomposite, achieving a linear response over a wide range of 3-1000 µM with fast response time of 5 s and sensitivity up to 2365.7 µAmM^{-1}cm^{-2}. The role of two hetero atoms as well as the method was studied by Di Geng & coauthors [110] who reported reduced graphene oxide/ nickel/ molybdenum disulfide nanoparticles composite using a deposition and precipitation method. The electrochemical measurements of the composite possessed a rapid response time of 2s, good reproducibility, stability and a wide linear range of 0.005 mM to 8.2 mM. Le Thuy Hoa & coauthors [111] proposed green synthesis using porous natured reduced graphene oxide/ silver nanoparticle composite, showing a fast response time of 11s showing sensitivity value of 725.0 µAcm^{-2}mM^{-1}. Chien-Te Hsieh & coworkers [112] used composite with improved catalytic performance for glucose detection with highest sensitivity upto 284 mAg^{-1}µM^{-1} along with fast response, wide linear range and selectivity. Graphene oxide doped polypyrrole/honeycomb like cobalt structure composite was introduced by Hui and Wang [113] via two-step electrochemical process, where the sensitivity was 297.73 µAmM^{-1}cm^{-2} along with rapid response time of 1s. Zhenyuan Ji and coauthors [114] fabricated electrode using graphene nanosheets with nickel nanoparticle via one-step thermal route. The electrochemical behavior towards glucose exhibited a sensitivity up to 388.4 µAmM^{-1}cm^{-2}, a wide linear range from 0.01-2.5 mM and detection limit of 0.79 µM. Jingyun Jiang & coauthors [115] developed a copper nanoparticle/ graphene nanoelectrode for electrochemical glucose detection which presented a detection limit of 0.12 µM, S/N=3 with a linear range up to 1 mM along with sensor response time <2 s. Natarajan Karikalan et al. [116] presented sulfur doped reduced graphene oxide supported copper sulfide nanoparticle composites using a sonication route, where the electrode allowed detection limit (32 nM) and a wide linear range (0.0001-3.88 mM and 3.88-20.17 mM) for the development of nonenzymatic glucose sensor. Shuang Li and coworkers [117] proposed a glucose sensor, consisted of reduced graphene oxide and 3-aminophenylboronic acid and it showed good sensitivity, selectivity, reproducibility and acceptable stability for the detection of glucose. Mozhdeh Mazaheri and coauthors [118] designed a electrode using reduced graphene oxide/nickel/zinc oxide heterostuctures through hydrothermal route and the resultant glucose sensor was much found to be better than many other electrodes. Alan Meng & coauthors [119] used a two-step electrodeposition process to obtain the cobalt sulfide/graphene/ semiconducting polymer composite for electrochemical measurement of glucose. The sensor displayed a sensitivity of 113.46 $\mu AmM^{-1}cm^{-2}$, a wide linear range from 0.2-1380 μM, low detection limit of 0.079 μM and response time of 3 s. Jian-De Xie & coauthors [120] developed a glucose sensor consisted of graphite spheres/graphene oxide sheets/palladium catalyst where, the sensor significantly enhanced the sensitivity in excess of 3.7 times, compared to palladium-graphite sphere catalyst within the linear range of 1-12 mM. Yen-Linh Thi Ngo et al. [121] constructed a three dimensional reduced graphene oxide/$NiMn_2O_4$ spinel binary nanostructure glucose sensor which exhibited excellent catalytic performance such as sensitivity (1310.8 $\mu A\ mM^{-1}cm^{-2}$), fast response time (<3.5 s) and a wide linear range (2 μM – 20 mM) due to the high specific surface area, synergetic effects between two components and enhanced electrical conductivity. Yen Linh Thi Ngo & coauthors [122] synthesized silver/nickel oxide/reduced graphene oxide sensor through hydrothermal pathway. The electrochemical detection of the composite towards glucose exhibited a sensitivity of 1869.4 $\mu AmM^{-1}cm^{-2}$, fast response time (<4 s) and a wide linear range upto 25 mM. A glucose sensor based on nitrogen doped graphene and copper-nickel nanoparticle was developed by Luba Shabnam et al. [123]. The sensitivity was 7143 $\mu AmM^{-1}cm^{-2}$ at low concentration (0.01μM^{-1}mM) and at high concentration (2-20 mM) was observed as 1030 $\mu AmM^{-1}cm^{-2}$. Yuqin Wang & coworkers [124] proposed reduced graphene oxide based Cu_2O hollow cubes electrode using refluxing and the glucose sensor has enhanced the sensitivity upto 813.713 $\mu Acm^{-2}mM^{-1}$ with big linear range from 0.01-9.0 mM and less detection limit of 0.44 μM, S/N=3 compared to bare Cu_2O based sensor.Mingjun Wang and coauthors [125] established 3D graphene/Cu_2O nanoparticle composites for enzyme free glucose sensor. The detection limit, response time, sensitivity and linear range of the sensor were 0.14 ± 0.01 μM, 1.6 s, (2.31±0.03) × 10^3 $\mu AmM^{-1}cm^{-2}$ and 0.48 μM - 1813 μM. Kong-Lin Wu and coworkers [126] decorated a nanocomposite of reduced graphene oxide/copper-nickel core shell and it showed high sensitivity of 780 $\mu AmM^{-1}cm^{-2}$, a wide linear range (0.001-41 mM) and low detection limit of 0.5 µM, S/N=3 for nonenzymatic sensing of glucose.

The modification of graphene with nickel plasma composite was established by Hao Wu and coworkers [127] which showed excellent glucose sensing with high sensitivity of 2213 µAmM^{-1}cm^{-2}, short response time of less than 1 s, low limit of detection (1 µM), long term stability, selectivity and reproducibility. Bei Xue et al. [128] designed an electrode composed by graphene/Co$_3$O$_4$/NiCo$_2$O$_4$ double shell nanocages and the nanocomposite had the sensitivity of 0.304 mAmM^{-1}cm^{-2}, linear response range of 0.01 mM to 3.52 mM and detection limit of 0.384 µM, S/N=3. Jianping Yang and coauthors [129] decorated functionalized graphene with cupric oxide and poly-(dimethyl diallyl ammonium chloride) and the electrode presented a high sensitivity of 4982.2 µAmM^{-1} cm^{-2} with a detection limit of 0.20 µM towards glucose.Jing Wang and coworkers [130] synthesized nitrogen doped graphene/CuO nanoparticle sensor via mild hydrothermal method and got sensitivity (223.1 µAmM^{-1}cm^{-2}), detection limit (2.7 µM) and linear range (0.01-6.75 mM). Liang Yang coauthors [131] established graphene nanosheets/Cu$_{2+1}$O/3D copper foam supported composite which was used as a potential sensor for glucose detection. Nitrogen doped reduced graphene oxide/nano needle like copper oxide nanocomposite was designed and the sensor showed linear range from 0.5-639 µM and lowest detectable concentration (0.01 µM) towards glucose detection [132]. Zhang and Liu [133] established cobalt oxide based graphene oxide nanocomposite to improve the sensing performance, achieving low detection limit of 0.15 µM, which was much lower than other cobalt oxide based glucose sensor. Yuehua Zhang et al. [134] constructed a modified electrode based on nitrogen doped reduced graphene oxide, nickel hydroxide and glassy carbon electrode, which displayed a high electrocatalytic activity towards oxidation of glucose. Haiyan Zhang et al. [135] prepared a glucose sensor using graphene oxide and nickel oxide nanosheets and the sensor detected in a limit of 0.18 µM, showing sensitive value of 1138 µA mM^{-1}cm^{-2}in the range of 1 µM - 0.4 mM. Chongjun Zhao and coworkers [136] fabricated a sensor with CuS/reduced graphene oxide/CuS nanocomposite, where the detection of limit was determined to be 0.5 µM with S/N ratio of 3.

In 2018, Elham Asadian & coauthors [137] presented a modified graphene nanoribbons/Ni-Co layered double hydroxide nanosheets/glassy carbon electrode for sensitive nonenzymatic detection of glucose. The electrode achieved sensitivity of 344 µAmM^{-1}cm^{-2} and wide dynamic linear range from 5 µM - 0.8 mM within a detection limit of 0.6 µM. Faranak Foroughi and coworkers [138] built up modified graphene/CuO nanoparticle composites which showed very response towards glucose at pH 13 within the concentration of 0.21 µM - 12 mM along with sensitivity of 700 µAmM^{-1}cm^{-2}. J.

Lavanya & coauthors [139] fabricated a glucose sensor based on graphene nanosheets/graphene nanoribbon/nickel nanoparticles and it displayed sensitivity up to 2.3 mA/mMcm^2, a wide linear range from 5 nM-5 mM and detection limit of 2.5 nM. Min Wang & coauthors [140] fabricated reduced graphene oxide/zinc oxide-cobalt oxide nanocomposites based sensor for glucose which showed a range of sensitivity of 168.7 μAmM^{-1}cm^{-2}with detection limit of 1.3 μM. Li Wang and coworkers [141] introduced an electrode constructed by reduced graphene oxide/copper oxide-cobalt nanostructures/biomass. It showed superior performance with high sensitivity of 802.86 μAmM^{-1}cm^{-2} along with suitable linear range 0.01-3.95 mM for electrochemical sensing of glucose. Xianming Fu et al. [142] built up a sensor based on reduced graphene oxide/ultrasmall platinum nanowire, showing good catalytic performances toward glucose determination including high sensitivity of 56.11 μA (mmol/L)$^{-1}$cm^{-2}, wide linear range (0.032-1.89 mmol/L) and low detection limit (4.6 μmol/L). Sudkate Chaiyo and coworkers [143] introduced a paper based sensor using graphene/cobalt phthalocyanine/ionic liquid nanocomposite, this sensor comprised a detection limit of 0.67 μM, S/N=3, a wide linear range from 0.01-1.3 mM and 1.3-5.0 mM for less and more concentration of glucose determination. Sorour Darvishi and coworkers [144] synthesized a gelatin methacryloyl sensor constructed by reduced graphene oxide and nickel nanoparticles. The sensor demonstrated good response to glucose oxidation at a low potential (0.47 V) with vastrange of detection (0.15 μM–10 mM), high sensitivity of 0.056 mAmM^{-1} and a detection limit of 5 nM. Nitrogen doped graphene encapsulated with nickel- cobalt nitride was later on prepared showing superior glucose sensing performance within the detection range of 50 nM, S/N=3 [145]. Zhe-Peng Deng & coworkers [146] fabricated a modified nickel-iron alloy/graphene oxide hybrid for electrochemical glucose sensing, which showed higher sensitivity towards glucose to that of commonly used nickel nanoparticles, while K. Justice Babu and coworkers [147] designed a reduced graphene oxide/3D dendrite copper-cobalt hybrid, which demonstrated excellent glucose detection properties with lower detection limit of 0.15 μM. Another reduced graphene oxide based phenoxyl-dextran composite was reported and the hybrid sensor achieved high sensitivity and detection limit of 0.02 μM with S/N ratio of 3 towards glucose oxidation [148]. Songyue Lin & coauthors [149] reported nonenzymatic sensor with copper nanoparticles with excellent catalytic activity towards glucose and displayed high sensitivity value of 1.518 mAmM^{-1}cm^{-2}. Feng Luan et al. [150] constructed enzyme free glucose sensor by ionic liquid functionalized graphene with Ni$_3$S$_2$ nanomaterials. The electrochemical properties showed significant electrocatalytic activity i.e. sensitivity of 25.343 μAμM^{-1}cm^{-1} with a linear range from 0-500 μM and detection maxima of 0.161 μM. Lian Ma & coworkers [151] firstly prepared nickel-platinum/reduced graphene oxide nanocomposite by a two-step reduction method.

The sensor showed good glucose detection performance with linear range of 10.008 mM and detection limit of 8 µM. Fengjuan Miao & coworkers [152] provided a graphene/zinc oxide nanoparticle composite, modified by nickel foam and the sensitivity and linear range of the sensor towards glucose detection were 1635.52 $\mu AmM^{-1}cm^{-2}$ and 50-1000 µmol, respectively.

Haleh Naeim & coworkers [153] synthesized a modified reduced graphene oxide/ionic liquid/nickel-palladium nanoparticle hybrid, which displayed enhanced current response compared to other nickel-palladium hybrid, while Yu Peng et al. [154] reported graphene fibers/gold nanosheets based fiber shaped microelectrode for detection of glucose and it exhibited excellent performance with detection limit (1.15 µM) and sensitivity (1045.9 $\mu AmM^{-1}cm^{-2}$). Raza and Krupanidhi [155] presented nitrogen enriched graphene oxide hybrid, which exhibited excellent glucose sensing performance such as lower detection limit of 0.66 µM, sensitivity up to 670 $\mu AmM^{-1}cm^{-2}$, long term stability and reproducibility. In the same year, Tugba Soganci & coworkers [156] designed copper nanoparticle/CVD graphene nanocomposite and used as a nonenzymatic sensor platform for the detection of glucose having detection limit of 7.2 µM. Graphene film/nafion/nickel nanoparticle decorated glucose biosensor was established by Parviz Sukhrobov and coauthors [157] which presented a higher and lower sensitivities of 3437.25 $\mu AmM^{-1}cm^{-2}$ and 2848.6 $\mu AmM^{-1}cm^{-2}$ along with detection limit of 0.6-0.8 µM, S/N=3. A.T. Ezhil Vilian & coworkers [158] prepared a reduced graphene oxide/Co$_3$O$_4$ hexagon structured nanocomposite which was optimized to show high sensitivity and a very low values of detection limit of 1.315 $mAmM^{-1}cm^{-2}$ and 0.4 µM, respectively. Da Xu and coworkers [159] designed reduced graphene oxide/Ag-CuO nanoparticles based sensor which showed an extremely linear response from 0.01-28 mM and sensitivity of 214.37 $\mu A\ mM^{-1}cm^{-2}$ with a 0.76 µM (S/N=0) detection range of +0.6 V. Reduced graphene oxide/platinum-palladium/persimmon tannin composite was introduced by Yewei Xue & coauthors [160] for glucose detection which exhibited high stability and fast response time of <3 s. Xiaoyi Yan and coworkers [161] synthesized a nanocomposite decorated with reduced graphene oxide and CuS nanoflakes and the sensor presented a wide linear range from 1-1000 µM, a low detection limit of 0.19 µM, a sensitivity of 53.5 µM/(cm^2 mM) and a fast response time of <6 s. Shaojun Yang et al. [162] developed a nanocomposite which displayed remarkable catalytic performance for glucose detection, including sensitivity (13.08 $\mu A\ mM^{-1}cm^{-2}$), wide linear range (1.04-17.44 mM) and a low detection limit (45.8 µM).

Three dimensional graphene decorated with Ni$_3$N nanoparticle sensor wa established as a glucose sensor, providing a wide linear range from 0.1-7645.3 µM with sensitivity up to 101.9 $\mu AmM^{-1}cm^{-2}$ along with fast response time (5 s) [163]. Guangchao Zang &

coauthors [164] presented a nanocomposite consisting of graphene oxide/copper nanowires/metal-organic frameworks, to exhibit good catalytic performances towards glucose determination with low detection limit, good reproducibility and accuracy. Jie Zhou et al. [165] demonstrated graphene oxide/Ni-Mn layered double hydroxide composite and it exhibited a remarkable sensitivity for glucose oxidation (839.2 μAmM^{-1} cm^{-2}). Recently, Rukiye Ayranci and coworkers [166] developed a sensor using reduced graphene oxide/monodisperse platinum-nickel through ultrasonic hydroxide assisted reduction method. The electrochemical analysis showed a glucose sensitivity up to 171.92 $\mu A/mMcm^2$, linear range of 0.02-5.0 mM and the detection limit of 6.3 μM. Xuerong Chen et al. [167] synthesized graphene/polyhedral cobalt based zeolite imidazole frame nanocomposite which showed better electrocatalytic activity towards glucose detection compared with the same ratio of a physical mixture of graphene and polyhedral cobalt based zeolite imidazole. A three dimensional nitrogen doped holey graphene hydrogel fabricated with $NiCo_2O_4$ nanoflowers based sensor through a one-pot hydrothermal method, where it possessedlinear range from 0.005-10.95 mM and low detection limit of 0.39 μM [168]. Weiwei Mao and coworkers [169] proposed same three dimensional graphene foam integrated with nickel hydroxide sensor and the sensor displayed an appreciable sensitivity of 2366 $\mu AmM^{-1}cm^{-2}$ and low detection range of 0.32 μM. A modified glucose sensor based on reduced graphene oxide/copper oxide nanoparticles was prepared by Sima Pourbeyram & coauthors [170] for high sensitive detection of glucose. Using this sensor, the sensitivity of 4760 (± 3.2) $\mu AmM^{-1}cm^{-2}$ and the detection limit of 0.32 μM for glucose was obtained. Kiattisak Promsuwan & coauthors [171] proposed an electrode modified with graphene nanoplatelets/multi walled carbon nanotubes/palladium nanoparticle. This electrode showed good catalytic activity for glucose oxidation including sensitivity (83.0 and 52.9 $\mu AmM^{-1}cm^{-2}$), wide linear range of detection (0.025-10 and 10-100 mM) and the detection limit was 0.008 mM. Very recent, a high sensing biosensor with nitrogen containing graphene with high sensing performance was established atdifferent pH. At pH 13, the sensitivity of 774.23 $\mu AmM^{-1}cm^{-2}$ and a linear range from 0.13 μM-14 mM was observed. Moreover, the sensitivity of 122.336 $\mu AmM^{-1}cm^{-2}$, detection limit of 14.52 μM and the linear range from 14.52 μM - u10 mM was determined at biological pH 7.4 [172]. Porous holey nitrogen doped graphene/Pt-Pd composite was designed by Abdulwahab Salah & coworkers [173] which showed direct oxidation activity towards the glucose, achieving wide linear range of 100-4000 μM, sensitivity of 52.526 $\mu AmM^{-1}cm^{-2}$, responding very fast i.e. <3s. Kritsada Samoson & coauthors [174] proposed a modified the electrode with graphene oxide/poly (acrylic acid)/palladium nanoparticle and it provided a detection limit of 22 μmol L^{-1}, a limit of quantitation of 76 μmol L^{-1} as well as superior sensitivity, selectivity and repeatability. Lili Xiao et al. [175] reported an electrode of nitrogen doped reduced graphene oxide with cobaltous phosphate and the electrode attributed superior catalytic activity for glucose oxidation at S/N=3 with sensitivity of 2307 µAmM⁻¹cm⁻². Yong Zhang & coworkers [176] fabricated nitrogen doped carbon/reduced graphene oxide nanocomposite through a simple wet chemical method. The sensor exhibited excellent electrocatalytic activity for determination of glucose such as sensitivity (3172 µAmM⁻¹cm⁻²), detection limit (0.34 µM at S/N=3) and linear detection range (0.5-10.0 µM). Yanyan Zhu and coworkers [177] by using hydrothermal method prepared a modified electrode decorated by graphene frameworks/ultrafine platinum nanoparticle. It displayed enhanced electrocatalytic activity towards glucose detection with achieving detection limit of 30 nM and linear responses in the range of 0.1µM-0.01 mM and 0.01 mM - 20 mM.

In order to compare the properties of functionalized and non-functionalized graphene based nanocomposites, the details of the work done by various researchers for nonenzymatic electrochemical detection of glucose in the recent few years have been summarized in Table-1. It clearly indicated that the graphene and graphene oxide based composites with metal oxides, noble metals, metal sulphides etc., plays a pivotal role for the construction of non enzymatic glucose sensor. Besides, the transition metals such as Ni, Co and Cu have been extensively used in nonenzymatic glucose sensors due to their distinguished electrocatalytic behavior towards glucose oxidation. However, it was observed that graphene along with these transition metals has achieved higher sensitivity with low detection limit as electrode materials as compared to graphene oxide based electrodes for the determination of glucose.

Table 1 *A comparison of properties of functionalized and non-functionalized graphene based composites for nonenzymatic electrochemical detection of glucose.*

Raw material	Electrode composition	Sensitivity[µA mM⁻¹cm⁻²/mAcm⁻²mM⁻¹/mAg⁻¹µM⁻¹]	Detection limit [µM/nM]	Linear range[µM/mM/µM-mM]	Ref.
Graphene	3D Graphene Frameworks/Co₃O₄	122.16	157 × 109 M, S/N=3	-	[98]
	Graphene/Bismuth oxychloride	1.878 and 127.2	-	500 - 10	[100]
	Graphene/NiCo₂O₄	-	-	0-0.14	[101]
	Graphene/3D copper oxide/copper oxide	3120	-	-	[105]
	Modified graphene sheets Ni nanoparticles	388.4	0.79	0.01-2.5	[114]

Cu nanoparticle/modified graphene	20.6	0.12, S/N=3	0.120	[115]
3D graphene/Ni/zinc oxide nanorod arrays	2030	0.15,S/N =3	0.5-1.11	[118]
Graphene/Cobalt sulfide/semi-onducting polymer	113.46	0.079	0.2-1380	[119]
3D graphene/Cu_2O nanoparticle	$(2.31\pm0.03) \times 103$	0.14 ± 0.01	0.48 -1813	[125]
Graphene functionalized PDDA /CuO	4982.2	0.20	-	[129]
Modified graphene with nickel plasma	2213	1	-	[127]
Graphene/Co_3O_4/$NiCo_2O_4$ double shell nanocages	0.304	0.384, S/N=3	0.01 - 3.52	[128]
Graphene nanosheets /$Cu_{2+1}O$/3D copper foam	3.076	5	-	[131]
Graphene nano ribbons/Ni-Co layered double hydroxide nanosheets/GCE	344	0.6	5-0.8	[137]
Modified graphene/CuO	700	-	-	[138]
Graphene nanosheets & nanoribbon/Ni nanoparticles	2.3	2.5	5-5	[139]
Graphene/cobalt-phthalocyanine/ionic liquid	-	0.67, S/N=3	0.01-1.3 & 1.3-5.0	[143]
Graphene/zinc oxide nanoparticle	1635.52	-	50-1000 μmol	[152]
Graphene fibers/gold nanosheets	1045.9	1.15	-	[154]
CVDgraphene/Cunanoparticle	430.52	7.2	0.01–1.0	[156]
Graphene/nafion/nickel nanoparticle	3437.25 & 2848.6	0.6-0.8,S/N= 3.	-	[157]
3D Graphene/ Ni_3N nanoparticle	101.9	-	0.1-7645.3	[163]
Graphene/polyhedral Co/zeolite imidazole	1521.1	0.36, S/N = 3	1–805.5	[167]
Graphene integrated with nickel hydroxide	2366	0.32	-	[169]
Graphene nanoplatelets /MWCNT/palladium	83.0 and 52.9	0.008	0.025-10 &10-100	[171]

	nanoparticle				
	Graphene/ultrafine platinum nanoparticle	46060.and205.23	30	0.1-0.01 & 0.01-20	[177]
Graphene oxide (GO)	GO- Imprinted Polymer	-	0.1	-	[106]
	GO/Co/doped polypyrrole	297.73	-	-	[113]
	GO/Cobalt oxide nanosheets	-	0.15	-	[133]
	GO/nickel oxide nanosheets	1138	0.18	1-0.4	[135]
	GO sheets /Graphite spheres/palladium	1.18, 0.54 & 0.32	-	1-12	[120]
	GO/Modified nickel-iron alloy	173	-	5	[146]
	GO/Ionic/liquid/Ni$_3$S$_2$	25.343	0.161	0-500	[150]
	GO/copper nanowires/metal-organic frameworks	-	7, S/N=3	20-26.6	[164]
	GO/Ni-Mn layered double hydroxide	839.2	1.2	-	[165]
	GO/poly(acrylic acid)/palladium nanoparticle	-	22 mumol L^{-1}	50-15,000& 15,000-60,000 mumolL^{-1}	[174]

Conclusions

The regular monitoring of the blood in the body is highly required to know the glucose concentration for reducing the risk of diabetes mellitus and other related diseases. In this context, the top issue is to search an glucose sensor, which is ideal in nature. Hence, the development of highly sensitive and interference free nonenzymatic glucose sensor is still greatly in demand. The electrochemical nonenzymatic sensors for glucose detection play a leading role in the sensing area, nowadays. This chapter approaches with the design, fabrication and development of different electrodes with functionalized graphene, heteroatom incorporated graphene, graphene derivatives based nanocomposites for sensing of glucose. It covers a detailed literature review on the sensitivity, detection limit and linear range of both functionalized and non-functionalized graphene with metals and polymers based electrodes towards the non enzymatic electrochemical sensing of the glucose molecules.

Acknowledgements

Authors are grateful to Director, NIT Agartala.

References

[1]   C. Shan, H. Yang, J. Song, D. Han, A. Ivaska, L. Niu, Direct electrochemistry of glucose oxidase and biosensing for glucose based on graphene. Anal Chem. 81(2009) 2378-2382.https://doi.org/10.1021/ac802193c

[2]   H. Gao, F. Xiao, C. B. Ching, H. Duan, One-step electrochemical synthesis of PtNi nanoparticle-graphene nanocomposites for nonenzymatic amperometric glucose detection, ACS appl mater inter. 3(2011) 3049-3057.https://doi.org/10.1021/am200563f

[3]   J. Y. Sun, K. J. Huang, Y. Fan, Z. W. Wu, D. D. Li, Glassy carbon electrode modified with a film composed of Ni(II), quercetin and graphene for enzyme-less sensing of glucose, Microchim Acta. 174(2011) 289.https://doi.org/10.1007/s00604-011-0625-0

[4]   Y. Qian, F. Ye, J. Xu, Z. G. Le, Synthesis of cuprous oxide (Cu_2O) nanoparticles/graphene composite with an excellent electrocatalytic activity towards glucose, Int. J. Electrochem. Sci. 7(2012) 10063-10073

[5]   X. C. Dong, H. Xu, X. W. Wang, Y. X. Huang, M. B. Chan-Park, H. Zhang, P. Chen, 3D graphene–cobalt oxide electrode for high-performance supercapacitor and enzymeless glucose detection, ACS nano, 6(2012) 3206-3213.https://doi.org/10.1021/nn300097q

[6]   Y. W. Hsu, T. K. Hsu, C. L. Sun, Y. T. Nien, N. W. Pu, M. D. Ger, Synthesis of CuO/graphene nanocomposites for nonenzymatic electrochemical glucose biosensor applications, Electrochim. Acta. 82(2012) 152-157.https://doi.org/10.1016/j.electacta.2012.03.094

[7]   F. Y. Kong, X. R. Li, W. W. Zhao, J. J. Xu, H. Y. Chen, Graphene oxide–thionine–Au nanostructure composites: preparation and applications in non-enzymatic glucose sensing, Electrochem. 14(2012)59-62.https://doi.org/10.1016/j.elecom.2011.11.004

[8]   C. Y. Lu, J. F. Xia, Z. H. Wang, Y. Z. Xia, F. F. Zhang, Nonenzymatic Electrochemical Detection of Glucose Using Well-Distributed Gold Nanoparticles on Graphene/Carbon Nanotube Nanohybrids, Adv. Mater. Res. 600 (2012) 234-237.https://doi.org/10.4028/www.scientific.net/AMR.600.234

[9]   L. Luo, L. Zhu, Z. Wang, Nonenzymatic amperometric determination of glucose by CuO nanocubes–graphene nanocomposite modified electrode, Bioelectrochemistry. 88 (2012)156-163.https://doi.org/10.1016/j.bioelechem.2012.03.006

[10] J. Luo, H. Zhang, S. Jiang, J. Jiang, X. Liu, Facile one-step electrochemical fabrication of a non-enzymatic glucose-selective glassy carbon electrode modified with copper nanoparticles and graphene, Microchim Acta. 177 (2012) 485-490.https://doi.org/ 10.1007/s00604-012-0795-4

[11] Z. Wang, Y. Hu, W. Yang, M. Zhou, X. Hu, Facile one-step microwave-assisted route towards Ni nanospheres/reduced graphene oxide hybrids for non-enzymatic glucose sensing, Sensors. 12(2012) 4860-4869.https://doi.org/10.3390/s120404860

[12] X. Wang, X. Dong, Y. Wen, C. Li, Q. Xiong, P. Chen, A graphene–cobalt oxide based needle electrode for non-enzymatic glucose detection in micro-droplets, Chem Commun. 48(2012) 6490-6492.https://doi.org/10.1039/c0xx00000x

[13] Y. Li, F. Huang, J. Chen, T. Mo, S. Li, F. Wang, Y. Li, A high performance enzyme-free glucose sensor based on the graphene-CuO nanocomposites, Int. J. Electrochem. Sci. 8 (2013) 6332-6342

[14] Z. Wang, J. Xia, X. Qiang, Y. Xia, G. Shi, F. Zhang, J. Tang, Polymer-assisted in situ growth of copper nanoparticles on graphene surface for non-enzymatic electrochemical sensing of glucose, Int. J. Electrochem. Sci. 8 (2013) 6941-6950

[15] Z. Fan, B. Liu, X. Liu, Z. Li, H. Wang, S. Yang, J. Wang, A flexible and disposable hybrid electrode based on Cu nanowires modified graphene transparent electrode for non-enzymatic glucose sensor, Electrochim Acta. 109 (2013) 602-608.https://doi.org/10.1016/j.electacta.2013.07.153

[16] M. F. Hossain, J. Y. Park, Palladium nanoparticles on electrochemically reduced chemically modified graphene oxide for non-enzymatic bimolecular sensing. Rsc Adv. 3(2013) 16109-16115.https://doi.org/ 10.1039/c3ra41235k

[17] M. Liu, R. Liu, W. Chen, Graphene wrapped Cu_2O nanocubes: non-enzymatic electrochemical sensors for the detection of glucose and hydrogen peroxide with enhanced stability, Biosens Bioelectron. 45(2013) 206-212.https://doi.org/ 10.1016/j.bios.2013.02.010

[18] W. Lu, X. Qin, A. M. Asiri, A. O. Al-Youbi, X. Sun, Facile synthesis of novel Ni (II)-based metal–organic coordination polymer nanoparticle/reduced graphene oxide nanocomposites and their application for highly sensitive and selective nonenzymatic glucose sensing, Analyst. 138(2013) 429-433.https://doi.org/10.1039/c2an36194a

[19] P. Si, X. C. Dong, P. Chen, D. H. Kim, A hierarchically structured composite of Mn 3 O 4/3D graphene foam for flexible nonenzymatic biosensors, J. Mater Chem B. 1(2013) 110-115.https://doi.org/10.1039/C2TB00073C

[20]   L. Wang, F. Nie, J. Zheng, Nickel Hydroxide and Intercalated Graphene with Ionic Liquid Nanocomposite-modified Electrode for Sensing of Glucose, J Chin Chem Soc. 60(2013) 1062-1069.https://doi.org/10.1002/jccs.201200587

[21]   D. Ye, G. Liang, H. Li, J. Luo, S. Zhang, H. Chen, J. Kong, A novel nonenzymatic sensor based on CuO nanoneedle/graphene/carbon nanofiber modified electrode for probing glucose in saliva. Talanta. 116 (2013) 223-230.https://doi.org/10.1016/j.talanta.2013.04.008

[22]   B. Yuan, C. Xu, L. Liu, Q. Zhang, S. Ji, L. Pi, Q. Huo, Cu_2O/NiO_x/graphene oxide modified glassy carbon electrode for the enhanced electrochemical oxidation of reduced glutathione and nonenzyme glucose sensor, Electrochim Acta. 104 (2013) 78-83.https://doi.org/10.1016/j.electacta.2013.04.073

[23]   H. Zhang, X. Xu, Y. Yin, P. Wu, C. Cai, Nonenzymatic electrochemical detection of glucose based on Pd_1Pt_3–graphene nanomaterials, J. Electroanal. Chem. 690 (2013) 19-24.https://doi.org/10.1016/j.jelechem.2012.12.001

[24]   Y. Zhang, X. Xiao, Y. Sun, Y. Shi, H. Dai, P. Ni, L. Wang, Electrochemical deposition of nickel nanoparticles on reduced graphene oxide film for nonenzymatic glucose sensing, Electroanal. 25(2013) 959-966.https://doi.org/10.1002/elan.201200479

[25]   S. Badhulika, R. K. Paul, T. Terse, A. Mulchandani, Nonenzymatic Glucose Sensor Based on Platinum Nanoflowers Decorated Multiwalled Carbon Nanotubes-Graphene Hybrid Electrode, Electroanal. 26(2014) 103-108.https://doi.org/10.1002/elan.201300286

[26]   X. Chen, X. Tian, L. Zhao, Z. Huang, M. Oyama, Nonenzymatic sensing of glucose at neutral pH values using a glassy carbon electrode modified with graphene nanosheets and Pt-Pd bimetallic nanocubes, Microchim Acta. 181(2014), 783-789.https://doi.org/10.1007/s00604-013-1142-0

[27]   Y. Zheng, P. Li, H. Li, S. Chen, Controllable growth of cobalt oxide nanoparticles on reduced graphene oxide and its application for highly sensitive glucose sensor, Int. J. Electrochem. Sci. 9 (2014) 7369-7381

[28]   T. Alizadeh, S. Mirzagholipur, A Nafion-free non-enzymatic amperometric glucose sensor based on copper oxide nanoparticles–graphene nanocomposite, Sensor Actuat B Chem. 198 (2014) 438-447.https://doi.org/10.1016/j.snb.2014.03.049.

[29]   S. Ci, S. Mao, T. Huang, Z. Wen, D. A. Steeber, J. Chen, Enzymeless glucose detection based on CoO/graphene microsphere hybrids, Electroanal. 26(2014) 1326-1334.doi.10.1002/elan.201300645

[30]   K. Dhara, J. Stanley, T. Ramachandran, B. G. Nair, S. B. TG, Pt-CuO nanoparticles decorated reduced graphene oxide for the fabrication of highly sensitive non-enzymatic disposable glucose sensor. Sensor Actuat B Chem. 195 (2014) 197-205.https://doi.org/10.1016/j.snb.2014.01.044

[31]   X. Feng, C. Guo, L. Mao, J. Ning, Y. Hu, Facile growth of Cu_2O nanowires on reduced graphene sheets with high nonenzymatic electrocatalytic activity toward glucose, J Am Ceram Soc. 97(2014) 811-815.https://doi.org/10.1111/jace.12686

[32]   W. Gao, W. W. Tjiu, J. Wei, T. Liu, Highly sensitive nonenzymatic glucose and H_2O_2 sensor based on $Ni(OH)_2$/electroreduced graphene oxide– Multiwalled carbon nanotube film modified glass carbon electrode, Talanta. 120(2014) 484-490.https://doi.org/10.1016/j.talanta.2013.12.012

[33]   H. U. Yao-Juan, D. U. Wen-Ji, C. H. E. N. Chang-Yun, Fabrication of flower-shaped Pt-Au-graphene nanostructure and its application in electrochemical detection of glucose, Chinese J Anal Chem. 42(2014) 1240-1244.https://doi.org/10.1016/S1872-2040(14)60764-7

[34]   Y. Hu, F. He, A. Ben, C. Chen, Synthesis of hollow Pt–Ni–graphene nanostructures for nonenzymatic glucose detection, Journal of Electroanal Chem. 726 (2014) 55-61.https://doi.org/10.1016/j.jelechem.2014.05.012

[35]   H. Huo, C. Guo, G. Li, X. Han, C. Xu, Reticular-vein-like Cu@ Cu_2O/reduced graphene oxide nanocomposites for a non-enzymatic glucose sensor, Rsc Adv. 4(2014) 20459-20465.https://doi.org/ 10.1039/c4ra02390k

[36]   N. S. Ismail, Q. H. Le, H. Yoshikawa, M. Saito, E. Tamiya, Development of non-enzymatic electrochemical glucose sensor based on graphene oxide nanoribbon–gold nanoparticle hybrid, Electrochim acta. 146 (2014) 98-105.https://doi.org/10.1016/j.electacta.2014.08.123

[37]   D. Jiang, Q. Liu, K. Wang, J. Qian, X. Dong, Z. Yang, B. Qiu, Enhanced non-enzymatic glucose sensing based on copper nanoparticles decorated nitrogen-doped graphene. Biosensors and Bioelectronics, 54 (2014) 273-278.https://doi.org/10.1016/j.bios.2013.11.005

[38]   Y. Kong, Y. Sha, Y. Tao, Y. Qin, H. Xue, M. Lu, Non-enzymatic glucose sensor based on nickel hexacyanoferrate/polyaniline hybrids on graphene prepared by a one-

step process, J Electrochem Soc. 161(2014) B269-B274.https://doi.org/ 10.1149/2.0961412jes

[39]   M. Li, X. Bo, Y. Zhang, C. Han, L. Guo, One-pot ionic liquid-assisted synthesis of highly dispersed PtPd nanoparticles/reduced graphene oxide composites for nonenzymatic glucose detection, Biosens Bioelectron. 56 (2014) 223-230.https://doi.org/10.1016/j.bios.2014.01.030

[40]   X. Lu, Y. Ye, Y. Xie, Y. Song, S. Chen, P. Li, L. Wang, Copper coralloid granule/polyaniline/reduced graphene oxide nanocomposites for nonenzymatic glucose detection. Anal Methods. 6(2014) 4643-4651.https://doi.org/ 10.1039/c4ay00421c

[41]   P. Subramanian, J. Niedziolka-Jonsson, A. Lesniewski, Q. Wang, M. Li, R. Boukherroub, S. Szunerits, Preparation of reduced graphene oxide–Ni (OH) 2 composites by electrophoretic deposition: application for non-enzymatic glucose sensing, J Mater Chem A. 2(2014) 5525-5533.https://doi.org/ 10.1039/C4TA00123K

[42]   L. Wang, Y. Zheng, X. Lu, Z. Li, L. Sun, Y. Song, Dendritic copper-cobalt nanostructures/reduced graphene oxide-chitosan modified glassy carbon electrode for glucose sensing, Sensor Actuat B Chem. 195 (2014) 1-7.https://doi.org/10.1016/j.snb.2014.01.007

[43]   X. Wang, E. Liu, X. Zhang, Non-enzymatic glucose biosensor based on copper oxide-reduced graphene oxide nanocomposites synthesized from water-isopropanol solution, Electrochim Acta. 130 (2014) 253-260.https://doi.org/10.1016/j.electacta.2014.03.030

[44]   B. Wang, S. Li, J. Liu, M. Yu, Preparation of nickel nanoparticle/graphene composites for non-enzymatic electrochemical glucose biosensor applications, Mater Res Bull. 49 (2014) 521-524.https://doi.org/10.1016/j.materresbull.2013.08.066

[45]   Y. Ye, P. Wang, E. Dai, J. Liu, Z. Tian, C. Liang, G. Shao, A novel reduction approach to fabricate quantum-sized SnO_2-conjugated reduced graphene oxide nanocomposites as non-enzymatic glucose sensors, Phys Chem Chem Phys. 16(2014) 8801-8807.https://doi.org/ 10.1039/c4cp00554f

[46]   B. Zhan, C. Liu, H. Chen, H. Shi, L. Wang, P. Chen, X. Dong, Free-standing electrochemical electrode based on $Ni(OH)_2$/3D graphene foam for nonenzymatic glucose detection, Nanoscale. 6(2014) 7424-7429.https://doi.org/ 10.1039/C4NR01611D

[47]   D. L. Zhou, J. J. Feng, L. Y. Cai, Q. X. Fang, J. R. Chen, A. J. Wang, Facile synthesis of monodisperse porous Cu2O nanospheres on reduced graphene oxide for

non-enzymatic amperometric glucose sensing, Electrochim Acta. 115 (2014) 103-108.https://doi.org/10.1016/j.electacta.2013.10.151

[48] P. Lu, J. Yu, Y. Lei, S. Lu, C. Wang, D. Liu, Q. Guo, Synthesis and characterization of nickel oxide hollow spheres–reduced graphene oxide–nafion composite and its biosensing for glucose, Sensor Actuat B-Chem. 208(2015) 90-98.https://doi.org/10.1016/j.snb.2014.10.140

[49] R. Aswathi, M. M. Ali, A. Shukla, K. Y. Sandhya, A green method to gold–graphene nanocomposite from cyclodextrin functionalized graphene for efficient non-enzymatic electrochemical sensing applications, Rsc Adv. 5(2015) 32027-32033.https://doi.org/ 10.1039/C4RA17323F

[50] J. Ding, W. Sun, G. Wei, Z. Su, Cuprous oxide microspheres on graphene nanosheets: an enhanced material for non-enzymatic electrochemical detection of H_2O_2 and glucose. Rsc Adv. 5(2015) 35338-35345.https://doi.org/ 10.1039/c5ra04164c

[51] S. Fu, G. Fan, L. Yang, F. Li, Non-enzymatic glucose sensor based on Au nanoparticles decorated ternary Ni-Al layered double hydroxide/single-walled carbon nanotubes/graphene nanocomposite, Electrochim acta. 152(2015) 146-154.https://doi.org/10.1016/j.electacta.2014.11.115

[52] N. Hui, S. Wang, H. Xie, S. Xu, S. Niu, X. Luo, Nickel nanoparticles modified conducting polymer composite of reduced graphene oxide doped poly (3,4-ethylenedioxythiophene) for enhanced nonenzymatic glucose sensing, Sensor Actuat B-Chem. 221(2015) 606-613.https://doi.org/10.1016/j.snb.2015.07.011

[53] N. Hui, W. Wang, G. Xu, X. Luo, Graphene oxide doped poly (3,4-ethylenedioxythiophene) modified with copper nanoparticles for high performance nonenzymatic sensing of glucose, J Mater Chem B. 3(2015) 556-561.https://doi.org/ 10.1039/c4tb01831a

[54] S. Hussain, K. Akbar, D. Vikraman, D. C. Choi, S. J. Kim, K. S. An, J. Jung, A highly sensitive enzymeless glucose sensor based on 3D graphene–Cu hybrid electrodes, New J Chem. 39(2015) 7481-7487.https://doi.org/10.1039/C5NJ01512J

[55] G. Li, H. Huo, C. Xu, $Ni_{0.31}Co_{0.69}S_2$ nanoparticles uniformly anchored on a porous reduced graphene oxide framework for a high-performance non-enzymatic glucose sensor, J Mater Chem A. 3(2015), 4922-4930.https://doi.org/ 10.1039/c4ta06553k

[56] Z. Liu, Y. Guo, C. Dong, A high performance nonenzymatic electrochemical glucose sensor based on polyvinylpyrrolidone–graphene nanosheets–nickel

nanoparticles–chitosan nanocomposite, Talanta. 137(2015) 87-93.https://doi.org/10.1016/j.talanta.2015.01.037

[57]   P. M. Nia, W. P. Meng, F. Lorestani, M. R. Mahmoudian, Y. Alias, Electrodeposition of copper oxide/polypyrrole/reduced graphene oxide as a nonenzymatic glucose biosensor, Sensor Actuat B-Chem 209 (2015) 100-108.https://doi.org/10.1016/j.snb.2014.11.072

[58]   S. Muralikrishna, K. Sureshkumar, Z. Yan, C. Fernandez, T. Ramakrishnappa, Non-enzymatic amperometric determination of glucose by CuO nanobelt graphene composite modified glassy carbon electrode, J Brazil Chem Soc. 26(2015) 1632-1641.https://doi.org/10.5935/0103-5053.20150134

[59]   S. Radhakrishnan, S. J. Kim, Facile fabrication of NiS and a reduced graphene oxide hybrid film for nonenzymatic detection of glucose, Rsc Adv. 5(2015) 44346-44352.https://doi.org/ 10.1039/C5RA01074H

[60]   I. Shackery, U. Patil, M. J. Song, J. S. Sohn, S. Kulkarni, S. Some, S. C. Jun, Sensitivity Enhancement in Nickel Hydroxide/3D-Graphene as Enzymeless Glucose Detection, Electroanal. 27(2015) 2363-2370.https://doi.org/10.1002/elan.201500009

[61]   L. Shahriary, A. A. Athawale, Electrochemical deposition of silver/silver oxide on reduced graphene oxide for glucose sensing, J Solid State Electr. 19(2015) 2255-2263.https://doi.org/ 10.1007/s10008-015-2865-0

[62]   H. Shu, G. Chang, J. Su, L. Cao, Q. Huang, Y. Zhang, Y. He, Single-step electrochemical deposition of high performance Au-graphene nanocomposites for nonenzymatic glucose sensing, Sensor Actuat B-Chem. 220 (2015) 331-339.https://doi.org/10.1016/j.snb.2015.05.094

[63]   F. Tehrani, L. Reiner, B. Bavarian, Rapid prototyping of a high sensitivity graphene based glucose sensor strip, PloS one. 10(2015) e0145036. https://doi.org/10.1371/journal.pone.0145036

[64]   Y. Tian, Y. Liu, W. P. Wang, X. Zhang, W. Peng. CuO nanoparticles on sulfur-doped graphene for nonenzymatic glucose sensing, Electrochim acta. 156(2015) 244-251.https://doi.org/10.1016/j.electacta.2015.01.016

[65]   Q. Wang, Q. Wang, M. Li, S. Szunerits, R. Boukherroub, Preparation of reduced graphene oxide/Cu nanoparticle composites through electrophoretic deposition: application for nonenzymatic glucose sensing, Rsc Adv. 5(2015), 15861-15869.https://doi.org/10.1039/c4ra14132f

[66]   L. Wang, X. Lu, C. Wen, Y. Xie, L. Miao, S. Chen, Y. Song, One-step synthesis of Pt–NiO nanoplate array/reduced graphene oxide nanocomposites for nonenzymatic glucose sensing, J Mater Chem A. 3(2015) 608-616.https://doi.org/10.1039/c4ta04724a

[67]   X. Yan, J. Yang, L. Ma, X. Tong, Y. Wang, G. Jin, X. Y. Guo, Size-controlled synthesis of Cu_2O nanoparticles on reduced graphene oxide sheets and their application as non-enzymatic glucose sensor materials, J Solid State Electr. 19(2015) 3195-3199.https://doi.org/10.1007/s10008-015-2911-y

[68]   S. Yang, G. Li, G. Wang, J. Zhao, X. Gao, L. Qu, Synthesis of Mn_3O_4 nanoparticles/nitrogen-doped graphene hybrid composite for nonenzymatic glucose sensor, Sensor Actuat B-Chem. 221 (2015) 172-178.https://doi.org/10.1016/j.snb.2015.06.110

[69]   S.Yang, L. Liu, G. Wang, G. Li, D. Deng, L. Qu, One-pot synthesis of Mn3O4 nanoparticles decorated with nitrogen-doped reduced graphene oxide for sensitive nonenzymatic glucose sensing, J Electroanal Chem.755(2015) 15-21.https://doi.org/10.1016/j.jelechem.2015.07.021

[70]   X. Zhang, Q. Liao, S. Liu, W. Xu, Y. Liu, Y. Zhang, CuNiO nanoparticles assembled on graphene as an effective platform for enzyme-free glucose sensing, Anal. Chim. Acta. 858(2015) 49-54.https://doi.org/10.1016/j.aca.2014.12.007

[71]   Y. Zhao, X. Bo, L. Guo, Highly exposed copper oxide supported on three-dimensional porous reduced graphene oxide for non-enzymatic detection of glucose, Electrochim Acta. 176 (2015) 1272-1279.https://doi.org/10.1016/j.electacta.2015.07.143

[72]   P. Nayak, S. P. Nair, S. Ramaprabhu, Enzyme-less and low-potential sensing of glucose using a glassy carbon electrode modified with palladium nanoparticles deposited on graphene-wrapped carbon nanotubes, Microchim Acta. 183(2016) 1055-1062.https://doi.org/10.1007/s00604-015-1729-8

[73]   R. Prasad, V. Ganesh, B. R. Bhat, Nickel-oxide multiwall carbon-nanotube/reduced graphene oxide a ternary composite for enzyme-free glucose sensing, Rsc Adv. 6(2016) 62491-62500.https://doi.org/ 10.1039/C6RA08708F

[74]   S. M. Naghib, M. Rahmanian, M. A. Keivan, S. Asiaei, O. Vahidi, Novel magnetic nanocomposites comprising reduced graphene oxide/Fe_3O_4/gelatin utilized in ultrasensitive non-enzymatic biosensing, Int. J. Electrochem. Sci. 11(2016)10256-10269.https://doi.org/ 10.20964/2016.12.29

[75]   M. M. Farid, L. Goudini, F. Piri, A. Zamani, F. Saadati, Molecular imprinting method for fabricating novel glucose sensor: Polyvinyl acetate electrode reinforced by MnO_2/CuO loaded on graphene oxide nanoparticles, Food chem. 194 (2016) 61-67.https://doi.org/10.1016/j.foodchem.2015.07.128

[76]   H. Heidari, E. Habibi, Amperometric enzyme-free glucose sensor based on the use of a reduced graphene oxide paste electrode modified with electrodeposited cobalt oxide nanoparticles, Microchim Acta. 183(2016) 2259-2266.https://doi.org/10.1007/s00604-016-1862-z

[77]   J. S. Chung, S. H. Hur, A highly sensitive enzyme-free glucose sensor based on Co_3O_4 nanoflowers and 3D graphene oxide hydrogel fabricated via hydrothermal synthesis, Sensor Actuat B-Chem. 223 (2016) 76-82.https://doi.org/10.1016/j.snb.2015.09.009

[78]   L. Ju, G. Wu, B. Lu, X. Li, H. Wu, A. Liu, Non-enzymatic Amperometric Glucose Sensor Based on Copper Nanowires Decorated Reduced Graphene Oxide, Electroanal. 28(2016) 2543-2551.https://doi.org/10.1002/elan.201600100

[79]   S. J. Li, L. L. Hou, B. Q. Yuan, M. Z. Chang, Y. Ma, J. M. Du, Enzyme-free glucose sensor using a glassy carbon electrode modified with reduced graphene oxide decorated with mixed copper and cobalt oxides, Microchim Acta. 183(2016), 1813-1821.https://doi.org/10.1007/s00604-016-1817-4

[80]   Y. Luo, F. Y. Kong, C. Li, J. J. Shi, W. X. Lv, W. Wang, One-pot preparation of reduced graphene oxide-carbon nanotube decorated with Au nanoparticles based on protein for non-enzymatic electrochemical sensing of glucose. Sensor Actuat B-Chem. 234 (2016) 625-632.https://doi.org/10.1016/j.snb.2016.05.046

[81]   G. Ma, M. Yang, C. Li, H. Tan, L. Deng, S. Xie, Y. Song, Preparation of spinel nickel-cobalt oxide nanowrinkles/reduced graphene oxide hybrid for nonenzymatic glucose detection at physiological level, Electrochim Acta. 220 (2016) 545-553.https://doi.org/10.1016/j.electacta.2016.10.163

[82]   V. Sahu, S. Grover, M. Sharma, A. Pandey, G. Singh, R. K. Sharma, CuO/Reduced graphene oxide nanocomposite for high performance non-enzymatic, cost effective glucose sensor, Sensor Lett. 14(2016) 1117-1122.https://doi.org/10.1166/sl.2016.3727

[83]   I. Shackery, U. Patil, A. Pezeshki, N. M. Shinde, S. Kang, S. Im, S. C. Jun, Copper hydroxide nanorods decorated porous graphene foam electrodes for non-enzymatic glucose sensing, Electrochim Acta. 191 (2016) 954-961.https://doi.org/10.1016/j.electacta.2016.01.047

185

[84] I. Shackery, U. Patil, A. Pezeshki, N. M. Shinde, S. Im, S. C. Jun, Enhanced non-enzymatic amperometric sensing of glucose using $Co(OH)_2$ nanorods deposited on a three dimensional graphene network as an electrode material, Microchim Acta. 183(2016) 2473-2479.https://doi.org/10.1007/s00604-016-1890-8

[85] Z. Shahnavaz, P. M. Woi, Y. Alias, Electrochemical sensing of glucose by reduced graphene oxide-zinc ferrospinels, Appl Surf Sci. 379 (2016) 156-162.https://doi.org/10.1016/j.apsusc.2016.04.061

[86] Z. Shen, W. Gao, P. Li, X. Wang, Q. Zheng, H. Wu, K. Ding, Highly sensitive nonenzymatic glucose sensor based on Nickel nanoparticle–attapulgite-reduced graphene oxide-modified glassy carbon electrode, Talanta. 159 (2016) 194-199.https://doi.org/10.1016/j.talanta.2016.06.016

[87] T. D.Thanh, J. Balamurugan, S. H. Lee, N. H. Kim, J. H. Lee, Effective seed-assisted synthesis of gold nanoparticles anchored nitrogen-doped graphene for electrochemical detection of glucose and dopamine, Biosens Bioelectron. 81(2016) 259-267.https://doi.org/10.1016/j.bios.2016.02.070

[88] C. Wang, Y. Sun, X. Yu, D. Ma, J. Zheng, P. Dou, X. Xu, Ag–Pt hollow nanoparticles anchored reduced graphene oxide composites for non-enzymatic glucose biosensor, J Mater Sci Mater El. 27(2016) 9370-9378.https://doi.org/10.1007/s10854-016-4979-2

[89] T. Yang, J. Xu, L. Lu, X. Zhu, Y. Gao, H. Xing, Z. Liu, Copper nanoparticle/graphene oxide/single wall carbon nanotube hybrid materials as electrochemical sensing platform for nonenzymatic glucose detection, J Electroanal Chem. 761(2016) 118-124.https://doi.org/10.1016/j.jelechem.2015.12.015

[90] S. N. A. M. Yazid, I. M. Isa, N. Hashim, Novel alkaline-reduced cuprous oxide/graphene nanocomposites for non-enzymatic amperometric glucose sensor application, Mater Sci Eng. C. 68(2016) 465-473.https://doi.org/10.1016/j.msec.2016.06.006

[91] G. Zeng, W. Li, S. Ci, J. Jia, Z. Wen, Highly dispersed NiO nanoparticles decorating graphene nanosheets for non-enzymatic glucose sensor and biofuel cell, Sci rep. 6 (2016) 36454.https://doi.org/ 10.1038/srep36454

[92] C. Zhang, Y. Zhang, Z. Miao, M. Ma, X. Du, J. Lin, Q. Chen, Dual-function amperometric sensors based on poly (diallydimethylammoniun chloride)-functionalized reduced graphene oxide/manganese dioxide/gold nanoparticles nanocomposite, Sensor Actuat B-Chem. 222(2016) 663-673.https://doi.org/10.1016/j.snb.2015.08.114

[93]   X. Zhang, Z. Zhang, Q. Liao, S. Liu, Z. Kang, Y. Zhang, Nonenzymatic glucose sensor based on in situ reduction of Ni/NiO-graphene nanocomposite, Sensors. 16(2016) 1791.https://doi.org/10.3390/s16111791

[94]   A. Zhao, Z. Zhang, P. Zhang, S. Xiao, L. Wang, Y. Dong, F. Xiao, 3D nanoporous gold scaffold supported on graphene paper: Freestanding and flexible electrode with high loading of ultrafine PtCo alloy nanoparticles for electrochemical glucose sensing, Anal chim acta. 938(2016) 63-71.https://doi.org/10.1016/j.aca.2016.08.013

[95]   J. Zheng, W. Zhang, Z. Lin, C. Wei, W. Yang, P. Dong, S. Hu, Microwave synthesis of 3D rambutan-like CuO and CuO/reduced graphene oxide modified electrodes for non-enzymatic glucose detection, J Mater Chem B. 4(2016) 1247-1253.https://doi.org/10.1039/C5TB02624E

[96]   W. Zheng, L. Hu, L. Y. S. Lee, K. Y. Wong, Copper nanoparticles/polyaniline/graphene composite as a highly sensitive electrochemical glucose sensor, J Electroanal Chem. 781(2016) 155-160.https://doi.org/10.1016/j.jelechem.2016.08.004

[97]   X. Zhuang, C. Tian, F. Luan, X. Wu, L. Chen, One-step electrochemical fabrication of a nickel oxide nanoparticle/polyaniline nanowire/graphene oxide hybrid on a glassy carbon electrode for use as a non-enzymatic glucose biosensor, Rsc Adv. 6(2016) 92541-92546.https://doi.org/10.1039/C6RA14970G

[98]   L. Bao, T. Li, S. Chen, C. Peng, L. Li, Q. Xu, W. Xu, 3D Graphene Frameworks/Co₃O₄ Composites Electrode for High-Performance Supercapacitor and Enzymeless Glucose Detection, Small. 13(2017) 1602077.https://doi.org/10.1002/smll.201602077

[99]   K. Ghanbari, F. Ahmadi, NiO hedgehog-like nanostructures/Au/polyaniline nanofibers/reduced graphene oxide nanocomposite with electrocatalytic activity for non-enzymatic detection of glucose, Anal biochem. 518(2017) 143-153.https://doi.org/10.1016/j.ab.2016.11.020

[100]   A. I. Gopalan, N. Muthuchamy, K. P. Lee, A novel bismuth oxychloride-graphene hybrid nanosheets based non-enzymatic photoelectrochemical glucose sensing platform for high performances, Biosens Bioelectron. 89 (2017), 352-360.https://doi.org/10.1016/j.bios.2016.07.017

[101]   T. H. Ko, S. Radhakrishnan, M. K. Seo, M. S. Khil, H. Y. Kim, B. S. Kim, A green and scalable dry synthesis of NiCo₂O₄/graphene nanohybrids for high-performance supercapacitor and enzymeless glucose biosensor applications, J Alloys Compd. 696(2017) 193-200.https://doi.org/10.1016/j.jallcom.2016.11.234

[102]   J. Feng, Y. Wang, Y. Hou, L. Li, Hierarchical structured $ZnFe_2O_4@$ RGO@ TiO_2 composite as powerful visible light catalyst for degradation of fulvic acid, J Nanopart Res.19(2017) 178.https://doi.org/10.1007/s11051-017-3842-6

[103]   L. Shabnam, S. N. Faisal, A. K. Roy, E. Haque, A. I. Minett, V. G. Gomes, Doped graphene/Cu nanocomposite: a high sensitivity non-enzymatic glucose sensor for food, Food chem. 221(2017)751-759.https://doi.org/10.1016/j.foodchem.2016.11.107.

[104]M. Sreejesh, S. Dhanush, F. Rossignol, H. S. Nagaraja, Microwave assisted synthesis of rGO/ZnO composites for non-enzymatic glucose sensing and supercapacitor applications, Ceram Int. 43(2017) 4895-4903.https://doi.org/10.1016/j.ceramint.2016.12.140

[105]   Z. Zhang, P. Pan, X. Liu, Z. Yang, J. Wei, Z. Wei, 3D-copper oxide and copper oxide/few-layer graphene with screen printed nanosheet assembly for ultrasensitive non-enzymatic glucose sensing, Mater Chem Phys. 187(2017) 28-38.https://doi.org/10.1016/j.matchemphys.2016.11.032

[106]   S. Alexander, P. Baraneedharan, S. Balasubrahmanyan, S. Ramaprabhu, Highly sensitive and selective non enzymatic electrochemical glucose sensors based on Graphene Oxide-Molecular Imprinted Polymer, Mater Sci Eng C. 78(2017) 124-129.https://doi.org/10.1016/j.msec.2017.04.045

[107]   P. Balasubramanian, M. Velmurugan, S. M. Chen, K. Y. Hwa, Optimized electrochemical synthesis of copper nanoparticles decorated reduced graphene oxide: Application for enzymeless determination of glucose in human blood, J Electroanal Chem. 807(2017) 128-136.https://doi.org/10.1016/j.jelechem.2017.11.042

[108]   S. Darvishi, M. Souissi, F. Karimzadeh, M. Kharaziha, R. Sahara, S. Ahadian, Ni nanoparticle-decorated reduced graphene oxide for non-enzymatic glucose sensing: an experimental and modeling study, Electrochim Acta. 240 (2017) 388-398.https://doi.org/10.1016/j.electacta.2017.04.086

[109]   S. Felix, P. Kollu, S. K. Jeong, A. N. Grace, A novel CuO–N-doped graphene nanocomposite-based hybrid electrode for the electrochemical detection of glucose, Appl Phys A. 123(2017) 620. https://doi.org/10.1007/s00339-017-1217-6

[110]   D. Geng, X. Bo, L. Guo, Ni-doped molybdenum disulfide nanoparticles anchored on reduced graphene oxide as novel electroactive material for a non-enzymatic glucose sensor, Sensor Actuat B-Chem. 244 (2017)131-141.https://doi.org/10.1016/j.snb.2016.12.122

[111] N. T. Y. Linh, J. S. Chung, S. H. Hur, Green synthesis of silver nanoparticle-decorated porous reduced graphene oxide for antibacterial non-enzymatic glucose sensors, Ionics. 23(2017) 1525-1532.https://doi.org/10.1007/s11581-016-1954-0

[112] C. T. Hsieh, W. H. Lin, Y. F. Chen, D. Y. Tzou, P. Q. Chen, R. S. Juang, Microwave synthesis of copper catalysts onto reduced graphene oxide sheets for non-enzymatic glucose oxidation, J Taiwan Inst Chem E. 71(2017) 77-83.https://doi.org/10.1016/j.jtice.2016.12.038

[113] N. Hui, J. Wang, Electrodeposited honeycomb-like cobalt nanostructures on graphene oxide doped polypyrrole nanocomposite for high performance enzymeless glucose sensing, J Electroanal Chem. 798(2017) 9-16.https://doi.org/10.1016/j.jelechem.2017.05.021

[114] Z. Ji, Y. Wang, Q. Yu, X. Shen, N. Li, H. Ma, J. Wang, One-step thermal synthesis of nickel nanoparticles modified graphene sheets for enzymeless glucose detection. J colloid interface sci. 506(2017) 678-684.https://doi.org/10.1016/j.jcis.2017.07.064

[115] J. Jiang, P. Zhang, Y. Liu, H. Luo, A novel non-enzymatic glucose sensor based on a Cu-nanoparticle-modified graphene edge nanoelectrode, Anal Methods. 9(2017) 2205-2210.https://doi.org/10.1039/C7AY00084G

[116] N. Karikalan, R. Karthik, S. M. Chen, C. Karuppiah, A. Elangovan, Sonochemical synthesis of sulfur doped reduced graphene oxide supported CuS nanoparticles for the non-enzymatic glucose sensor applications, Sci rep. 7(2017) 2494.https://doi.org/10.1038/s41598-017-02479-5

[117] S. Li, Q. Zhang, Y. Lu, D. Ji, D. Zhang, J. Wu, Q. Liu, One step electrochemical deposition and reduction of graphene oxide on screen printed electrodes for impedance detection of glucose, Sensor Actuat B-Chem. 244(2017) 290-298.https://doi.org/10.1016/j.snb.2016.12.142

[118] M. Mazaheri, H. Aashuri, A. Simchi, Three-dimensional hybrid graphene/nickel electrodes on zinc oxide nanorod arrays as non-enzymatic glucose biosensors, Sensor Actuat B-Chem. 251 (2017) 462-471.https://doi.org/10.1016/j.snb.2017.05.062

[119] A. Meng, L. Sheng, K. Zhao, Z. Li, A controllable honeycomb-like amorphous cobalt sulfide architecture directly grown on the reduced graphene oxide–poly (3, 4-ethylenedioxythiophene) composite through electrodeposition for non-enzyme glucose sensing, J Mater Chem B. 5(2017) 8934-8943.https://doi.org/10.1039/C7TB02482G

[120] J. D. Xie, S. Gu, H. Zhang, Microwave deposition of palladium catalysts on graphite spheres and reduced graphene oxide sheets for electrochemical glucose Sensing, Sensors. 17(2017) 2163.https://doi.org/10.3390/s17102163

[121] Y. L. T. Ngo, L. Sui, W. Ahn, J. S. Chung, S. H. Hur, $NiMn_2O_4$ spinel binary nanostructure decorated on three-dimensional reduced graphene oxide hydrogel for bifunctional materials in non-enzymatic glucose sensor, Nanoscale. 9(2017) 19318-19327.https://doi.org/10.1039/C7NR07748C

[122] Y. L. T. Ngo, J. S. Chung, S. H. Hur, Multi-dimensional Ag/NiO/reduced graphene oxide nanostructures for a highly sensitive non-enzymatic glucose sensor. J Alloys Compd. 712(2017) 742-751.https://doi.org/10.1016/j.jallcom.2017.04.131

[123] L. Shabnam, S. N. Faisal, A. K. Roy, A. I. Minett, V. G. Gomes, Nonenzymatic multispecies sensor based on Cu-Ni nanoparticle dispersion on doped graphene, Electrochim Acta. 224(2017) 295-305.https://doi.org/10.1016/j.electacta.2016.12.056

[124] Y. Wang, Z. Ji, X. Shen, G. Zhu, J. Wang, X. Yue, Facile growth of Cu_2O hollow cubes on reduced graphene oxide with remarkable electrocatalytic performance for non-enzymatic glucose detection, New J Chem. 41(2017) 9223-9229.https://doi.org/10.1039/C7NJ01952A

[125] M. Wang, X. Song, B. Song, J. Liu, C. Hu, D. Wei, C. P. Wong, Precisely quantified catalyst based on in situ growth of Cu2O nanoparticles on a graphene 3D network for highly sensitive glucose sensor, Sensor Actuat B-Chem. 250(2017) 333-341.https://doi.org/10.1016/j.snb.2017.04.125

[126] K. L. Wu, Y. M. Cai, B. B. Jiang, W. C. Cheong, X. W. Wei, W. Wang, N. Yu, Cu@Ni core–shell nanoparticles/reduced graphene oxide nanocomposites for nonenzymatic glucose sensor, Rsc Adv. 7(2017) 21128-21135.https://doi.org/10.1039/C7RA00910K

[127] H. Wu, Y. Yu, W. Gao, A. Gao, A. M. Qasim, F. Zhang, P. K. Chu, Nickel plasma modification of graphene for high-performance non-enzymatic glucose sensing. Sensor Actuat B-Chem. 251(2017) 842-850.https://doi.org/10.1016/j.snb.2017.05.128

[128] B. Xue, K. Li, L. Feng, J. Lu, L. Zhang, Graphene wrapped porous Co_3O_4/NiCo2O4 double-shelled nanocages with enhanced electrocatalytic performance for glucose sensor, Electrochim Acta. 239(2017) 36-44.https://doi.org/10.1016/j.electacta.2017.04.005

[129] J. Yang, Q. Lin, W. Yin, T. Jiang, D. Zhao, L. Jiang, A novel nonenzymatic glucose sensor based on functionalized PDDA-graphene/CuO nanocomposites. Sensor Actuat B-Chem. 253(2017) 1087-1095.https://doi.org/10.1016/j.snb.2017.07.008

[130] J. Yang, W. Tan, C. Chen, Y. Tao, Y. Qin, Y. Kong, Nonenzymatic glucose sensing by CuO nanoparticles decorated nitrogen-doped graphene aerogel, Mater. Sci. Eng C. 78(2017) 210-217.https://doi.org/10.1016/j.msec.2017.04.097

[131] L. Yang, D. Liu, G. Cui, Y. Xie, $Cu_{2+1}O$/graphene nanosheets supported on three dimensional copper foam for sensitive and efficient non-enzymatic detection of glucose, Rsc Adv. 7(2017) 19312-19317.https://doi.org/10.1039/C7RA02011B

[132] S. Yang, G. Li, D. Wang, Z. Qiao, L. Qu, Synthesis of nanoneedle-like copper oxide on N-doped reduced graphene oxide: a three-dimensional hybrid for nonenzymatic glucose sensor, Sensor Actuat B-Chem.238(2017) 588-595.https://doi.org/10.1016/j.snb.2016.07.105

[133] H. Zhang, S. Liu, A combined self-assembly and calcination method for preparation of nanoparticles-assembled cobalt oxide nanosheets using graphene oxide as template and their application for non-enzymatic glucose biosensing, J colloid interf sci. 485(2017) 159-166.https://doi.org/10.1016/j.jcis.2016.09.041

[134] Y. Zhang, W. Lei, Q. Wu, X. Xia, Q. Hao, Amperometric nonenzymatic determination of glucose via a glassy carbon electrode modified with nickel hydroxide and N-doped reduced graphene oxide, Microchim Acta. 184(2017) 3103-3111.https://doi.org/10.1007/s00604-017-2332-y

[135] H. Zhang, S. Liu, Nanoparticles-assembled NiO nanosheets templated by graphene oxide film for highly sensitive non-enzymatic glucose sensing, Sensor Actuat B- Chem. 238(2017) 788-794.https://doi.org/10.1016/j.snb.2016.07.126

[136] C. Zhao, X. Wu, X. Zhang, P. Li, X. Qian, Facile synthesis of layered CuS/RGO/CuS nanocomposite on Cu foam for ultrasensitive nonenzymatic detection of glucose, J Electroanal Chem. 785(2017) 172-179.https://doi.org/10.1016/j.jelechem.2016.12.039

[137] E. Asadian, S. Shahrokhian, A. I. Zad, Highly sensitive nonenzymetic glucose sensing platform based on MOF-derived NiCo LDH nanosheets/graphene nanoribbons composite, J Electroanal Chem. 808(2018) 114-123.https://doi.org/10.1016/j.jelechem.2017.10.060

[138] F. Foroughi, M. Rahsepar, M. J. Hadianfard, H. Kim, Microwave-assisted synthesis of graphene modified CuO nanoparticles for voltammetric enzyme-free

sensing of glucose at biological pH values, Microchim Acta. 185(2018) 57.https://doi.org/10.1007/s00604-017-2558-8

[139] L. Jothi, N. Jayakumar, S. K. Jaganathan, G. Nageswaran, Ultrasensitive and selective non-enzymatic electrochemical glucose sensor based on hybrid material of graphene nanosheets/graphene nanoribbons/nickel nanoparticle, Mater Res Bull. 98(2018) 300-307.https://doi.org/ 10.1016/j.materresbull.2017.10.020

[140] M. Wang, J. Ma, Q. Chang, X. Fan, G. Zhang, F. Zhang, Y. Li, Fabrication of a novel ZnO–CoO/rGO nanocomposite for nonenzymatic detection of glucose and hydrogen peroxide, Ceram Int. 44(2018) 5250-5256.https://doi.org/10.1016/j.ceramint.2017.12.136

[141] L. Wang, L. Xu, Y. Zhang, H. Yang, L. Miao, C. Peng, Y. Song, Copper Oxide– Cobalt Nanostructures/Reduced Graphene Oxide/Biomass-Derived Macroporous Carbon for Glucose Sensing, ChemElectroChem. 5(2018) 501-506.https://doi.org/10.1002/celc.201701062

[142] X. Fu, Z. Chen, S. Shen, L. Xu, Z. Luo, Highly Sensitive Nonenzymatic Glucose Sensor Based on Reduced Graphene Oxide/Ultrasmall Pt Nanowire Nanocomposites, Int. J. Electrochem. 13(2018) 4817-4826.https://doi.org/10.20964/2018.05.46

[143] S. Chaiyo, E. Mehmeti, W. Siangproh, T. L. Hoang, H. P. Nguyen, O. Chailapakul, K. Kalcher, Non-enzymatic electrochemical detection of glucose with a disposable paper-based sensor using a cobalt phthalocyanine–ionic liquid–graphene composite, Biosens. Bioelectron. 102(2018) 113-120.https://doi.org/10.1016/j.bios.2017.11.015

[144] S. Darvishi, M. Souissi, M. Karaziha, F. Karimzadeh, R. Sahara, S. Ahadian, Gelatin methacryloyl hydrogel for glucose biosensing using Ni nanoparticles-reduced graphene oxide: An experimental and modeling study, Electrochim Acta. 261(2018) 275-283.https://doi.org/10.1016/j.electacta.2017.12.126

[145] T. Deepalakshmi, D. T. Tran, N. H. Kim, K. T. Chong, J. H. Lee, Nitrogen-doped graphene-encapsulated nickel cobalt nitride as a highly sensitive and selective electrode for glucose and hydrogen peroxide sensing applications, ACS Appl Mater Inter. 10(2018) 35847-35858.https://doi.org/10.1021/acsami.8b15069

[146] Z. P. Deng, Y. Sun, Y. C. Wang, J. D. Gao, A NiFe Alloy Reduced on Graphene Oxide for Electrochemical Nonenzymatic Glucose Sensing, Sensors. 18(2018) 3972.https://doi.org/10.3390/s18113972

[147] K. Justice Babu, S. Sheet, Y. S. Lee, G. Gnana Kumar, Three-Dimensional Dendrite Cu–Co/Reduced Graphene Oxide Architectures on a Disposable Pencil Graphite Electrode as an Electrochemical Sensor for Nonenzymatic Glucose Detection, ACS Sustain Chem Eng. 6(2018) 1909-1918.https://doi.org/10.1021/acssuschemeng.7b03314

[148] B. Li, A. Yu, G. Lai, Self-assembly of phenoxyl-dextran on electrochemically reduced graphene oxide for nonenzymatic biosensing of glucose, Carbon. 127(2018) 202-208.https://doi.org/10.1016/j.carbon.2017.10.096

[149] S. Lin, W. Feng, X. Miao, X. Zhang, S. Chen, Y. Chen, Y. Zhang, A flexible and highly sensitive nonenzymatic glucose sensor based on DVD-laser scribed graphene substrate, Biosens Bioelectron. 110(2018) 89-96.https://doi.org/10.1016/j.bios.2018.03.019

[150] F. Luan, S. Zhang, D. Chen, F. Wei, X. Zhuang, Ni_3S_2/ionic liquid-functionalized graphene as an enhanced material for the nonenzymatic detection of glucose, Microchem J. 143(2018) 450-456.https://doi.org/10.1016/j.microc.2018.08.046

[151] L. Ma, X. Wang, Q. Zhang, X. Tong, Y. Zhang, Z. Li, Pt catalyzed formation of a Ni@ Pt/reduced graphene oxide nanocomposite: preparation and electrochemical sensing application for glucose detection, Anal methods. 10(2018) 3845-3850. https://doi.org/10.1039/C8AY01275J

[152] F. Miao, F. Wu, R. Miao, W. Cong, Y. Zang, B. Tao, Graphene/nano-ZnO hybrid materials modify Ni-foam for high-performance electrochemical glucose sensors, Ionics. 24(2018) 4005-4014.https://doi.org/10.1007/s11581-018-2539-x

[153] H. Naeim, F. Kheiri, M. Sirousazar, A. Afghan, Ionic liquid/reduced graphene oxide/nickel-palladium nanoparticle hybrid synthesized for non-enzymatic electrochemical glucose sensing, Electrochim Acta. 282(2018) 137-146.https://doi.org/10.1016/j.electacta.2018.05.204

[154] Y. Peng, D. Lin, J. J. Gooding, Y. Xue, L. Dai, Flexible fiber-shaped non-enzymatic sensors with a graphene-metal heterostructure based on graphene fibres decorated with gold nanosheets, Carbon. 136(2018) 329-336.https://doi.org/10.1016/j.carbon.2018.05.004

[155] W. Raza, S. B. Krupanidhi, Engineering Defects in Graphene Oxide for Selective Ammonia and Enzyme-Free Glucose Sensing and Excellent Catalytic Performance for para-Nitrophenol Reduction, ACS appl mater inter. 10(2018) 25285-25294.https://doi.org/10.1021/acsami.8b05162

[156] T. Soganci, R. Ayranci, E. Harputlu, K. Ocakoglu, M. Acet, M. Farle, M. Ak, An effective non-enzymatic biosensor platform based on copper nanoparticles decorated by sputtering on CVD graphene, Sensor Actuat B-Chem. 273(2018) 1501-1507.https://doi.org/10.1016/j.snb.2018.07.064

[157] P. Sukhrobov, S. Numonov, S. Gao, X. Mamat, T. Wagberg, Y. Guo, G. Hu, Nonenzymatic Glucose Biosensor Based on NiNPs/Nafion/Graphene Film for Direct Glucose Determination in Human Serum, Nano. 13(2018) 1850075.https://doi.org/10.1142/S1793292018500753

[158] A. E. Vilian, B. Dinesh, M. Rethinasabapathy, S. K. Hwang, C. S. Jin, Y. S. Huh, Y. K. Han, Hexagonal Co 3 O 4 anchored reduced graphene oxide sheets for high-performance supercapacitors and non-enzymatic glucose sensing, J Mater Chem A. 6(2018) 14367-14379.https://doi.org/10.1039/C8TA04941F

[159] D. Xu, C. Zhu, X. Meng, Z. Chen, Y. Li, D. Zhang, S. Zhu, Design and fabrication of Ag-CuO nanoparticles on reduced graphene oxide for nonenzymatic detection of glucose. Sensor Actuat B-Chem 265(2018) 435-442.https://doi.org/10.1016/j.snb.2018.03.086

[160] Y. Xue, Y. Huang, Z. Zhou, G. Li, Non-enzymatic glucose biosensor based on reduction graphene oxide-persimmon tannin-Pt-Pd nanocomposite. In IOP Conference Series: Mater Sci Eng. 382(2018) 022016. https://doi.org/10.1088/1757-899X/382/2/022016

[161] X. Yan, Y. Gu, C. Li, B. Zheng, Y. Li, T. Zhang, M. Yang, A non-enzymatic glucose sensor based on the CuS nanoflakes–reduced graphene oxide nanocomposite, Anal methods. 10(2018) 381-388.https://doi.org/10.1039/C7AY02290E

[162] S.Yang, D. Liu, Q. B. Meng, S. Wu, X. M. Song, Reduced graphene oxide-supported methylene blue nanocomposite as a glucose oxidase-mimetic for electrochemical glucose sensing, Rsc adv. 8(2018) 32565-32573.https://doi.org/10.1039/C8RA06208K

[163] D. Yin, X. Bo, J. Liu, L. Guo, A novel enzyme-free glucose and H_2O_2 sensor based on 3D graphene aerogels decorated with Ni3N nanoparticles, Anal chim acta. 1038(2018) 11-20.https://doi.org/10.1016/j.aca.2018.06.086

[164] G. Zang, W. Hao, X. Li, S. Huang, J. Gan, Z. Luo, Y. Zhang, Copper nanowires-MOFs-graphene oxide hybrid nanocomposite targeting glucose electro-oxidation in neutral medium, Electrochim Acta. 277(2018) 176-184.https://doi.org/10.1016/j.electacta.2018.05.016

[165] J. Zhou, M. Min, Y. Liu, J. Tang, W. Tang, Layered assembly of NiMn-layered double hydroxide on graphene oxide for enhanced non-enzymatic sugars and hydrogen peroxide detection, Sensor Actuat B-Chem. 260(2018) 408-417.https://doi.org/10.1016/j.snb.2018.01.072

[166] R. Ayranci, B. Demirkan, B. Sen, A. Şavk, M. Ak, F. Şen, Use of the monodisperse Pt/Ni@ rGO nanocomposite synthesized by ultrasonic hydroxide assisted reduction method in electrochemical nonenzymatic glucose detection. Mater Sci Eng C 99(2019) 951-956.https://doi.org/10.1016/j.msec.2019.02.040

[167] X. Chen, D. Liu, G. Cao, Y. Tang, C. Wu, In Situ Synthesis of a Sandwich-like Graphene@ ZIF-67 Heterostructure for Highly Sensitive Nonenzymatic Glucose Sensing in Human Serums, ACS appl mater inter. 11(2019) 9374-9384.https://doi.org/10.1021/acsami.8b22478

[168] Z. Lu, L. Wu, J. Zhang, W. Dai, G. Mo, J. Ye, Bifunctional and highly sensitive electrochemical non-enzymatic glucose and hydrogen peroxide biosensor based on NiCo2O4 nanoflowers decorated 3D nitrogen doped holey graphene hydrogel, Mater Sci Eng C. 102(2019) 708-717.https://doi.org/10.1016/j.msec.2019.04.072

[169] W. Mao, H. He, Z. Ye, J. Huang, Three-dimensional graphene foam integrated with Ni(OH)2 nanosheets as a hierarchical structure for non-enzymatic glucose sensing. J Electroanal Chem. 832(2019) 275-283.https://doi.org/10.1016/j.jelechem.2018.11.016

[170] S. Pourbeyram, J. Abdollahpour, M. Soltanpour, Green synthesis of copper oxide nanoparticles decorated reduced graphene oxide for high sensitive detection of glucose, Mater Sci Eng C. 94(2019) 850-857.https://doi.org/10.1016/j.msec.2018.10.034

[171] K. Promsuwan, N. Kachatong, W. Limbut, Simple flow injection system for non-enzymatic glucose sensing based on an electrode modified with palladium nanoparticles-graphene nanoplatelets/mullti-walled carbon nanotubes, Electrochim Acta. 320(2019) 134621.https://doi.org/10.1016/j.electacta.2019.134621

[172] M. Rahsepar, F. Foroughi, H. Kim, A new enzyme-free biosensor based on nitrogen-doped graphene with high sensing performance for electrochemical detection of glucose at biological pH value, Sensor Actuat B-Chem. 282(2019) 322-330.https://doi.org/10.1016/j.snb.2018.11.078

[173] A. Salah, N. Al-Ansi, S. Adlat, M. Bawa, Y. He, X. Bo, L. Guo, Sensitive nonenzymatic detection of glucose at PtPd/porous holey nitrogen-doped graphene, J Alloys Compd. 792(2019) 50-58.https://doi.org/10.1016/j.jallcom.2019.04.021

[174]   K. Samoson, P. Thavarungkul, P. Kanatharana, W. Limbut, A Nonenzymatic Glucose Sensor Based on the Excellent Dispersion of a Graphene Oxide-Poly (acrylic acid)-Palladium Nanoparticle-Modified Screen-Printed Carbon Electrode, J Electrochem Soc. 166(2019) B1079-B1087. https://doi.org/10.1149/2.1381912jes

[175]   L. Xiao, Q. Chen, L. Jia, Q. Zhao, J. Jiang, Networked cobaltous phosphate decorated with nitrogen-doped reduced graphene oxide for non-enzymatic glucose sensing, Sensor Actuat B-Chem. 283(2019) 443-450.https://doi.org/10.1016/j.snb.2018.12.014

[176]   Y. Zhang, Y. Zhang, H. Zhu, S. Li, C. Jiang, R. J. Blue, Y. Su, Functionalization of the support material based on N-doped carbon-reduced graphene oxide and its influence on the non-enzymatic detection of glucose, J Alloys Compd. 780(2019) 98-106.https://doi.org/10.1016/j.jallcom.2018.11.368

[177]   Y. Zhu, X. Zhang, J. Sun, M. Li, Y. Lin, K. Kang, J. Wang, A non-enzymatic amperometric glucose sensor based on the use of graphene frameworks-promoted ultrafine platinum nanoparticles, Microchim Acta. 186(2019) 538.https://doi.org/10.1007/s00604-019-3653-9

Toxic Gas Sensors and Biosensors   Materials Research Forum LLC
Materials Research Foundations **92** (2021) 197-219 https://doi.org/10.21741/9781644901175-7

Chapter 7

Toxins and Pollutants Detection on Biosensor Surfaces

Amal I. Hassan, Hosam M. Saleh*

Radioisotope Department, Nuclear Research Center, Atomic Energy Authority, Dokki 12311, Giza, Egypt

virtualaml@gmail.com, *hosam.saleh@eaea.org.eg, *hosamsaleh70@yahoo.com

Abstract

Detecting toxins and pollutants is a field of biological sensing investigation. A rapid responsive and reliable biosensor has become an urgent requirement to help detect pathogenic bacteria and toxins that cause dangerous diseases. Recently studies have developed a multi-channel surface plasma resonance sensor for the simultaneous quantitative disclosure of food-borne bacterial pathogens. Now, biosensors are a dominant alternative to traditional analytic procedures to control both natural water quality and the water treatment used by the food manufacturing during the production method, and wastewater before it is released into natural waterways. The most significant attributes of biosensors are the immense sensitivity, short time interval, precision, and comparatively insignificant cost. Biometric sensors will discover the existence or coexist the content of varied cytotoxic materials (insecticides, serious metals, etc.) in both water and food. The existence of pollutants, particularly toxic metal ions within the water used in food manufacture processes, may be a potential field for biosensor applications. This chapter summarizes the evolution and application of biosensors to regulate and discover toxins and different pollutants. It will highlight the various biosensors and the future sight of this field.

Keywords

Biosensor, Toxins, Pollutants, Pathogens, Food-Borne

Contents

1. Introduction

Biosensors are a component or biological detection device that consists of the interference of whole bacteria or some of their products such as enzymes and antibodies with an electronic device to produce a measurable signal [1]. Electronic signals have many measurable qualities such as density and spectral properties, and energy is also measurable through several moving electronic units. This biological detector produces the interaction between the biological material and the reaction material to be measured, which changes into a pulse or an electronic or electrical signal utilizing a suitable energy transformer [2]. As well as, the design of vital reagents devices to sense change and respond to it, and the electrical signal is amplified in the biological sense device to give something that can be read on a digital screen or printer, and there are many types of changes that may occur that maybe the release of heat or light or a change in the pH or mass or the production of a new chemical compound [3]. Biological reagents have multiple shapes and sizes that are commonly used to monitor changes in environmental conditions [4]. With biological reagents, specific bacterial groupings and dangerous chemicals, or measure acidity levels can be discovered and measured or simply bacteria can be used to detect bacteria at the same time [4].

Biosensors can detect fine particles to help monitor human health and the environment. Scientists are developing sensors susceptible enough to recognize the smallest particle, which empowers us to reveal diseases or serious environmental pollutants [5].

A biological sensing system is an integrated device consisting of biological elements for identification, capable of monitoring biological interaction and converting it into a signal that can be processed [6]. It comprises 3 elements as shown in fig. 1. The susceptible biological aspects can be tissues, microorganisms, organelles, cell receptors, enzymes, antibodies, and nucleic acids, which is realized biologically or as a result of the excitation of those biological elements by any external effects to the energy electrical device or the detector; it works physically; as an optical piezoelectric and electrochemical) that converts the signals ensuing from the interaction of substitution with the biological component into another indicator (energy converters) where it becomes easy to measure and regulate them [7].

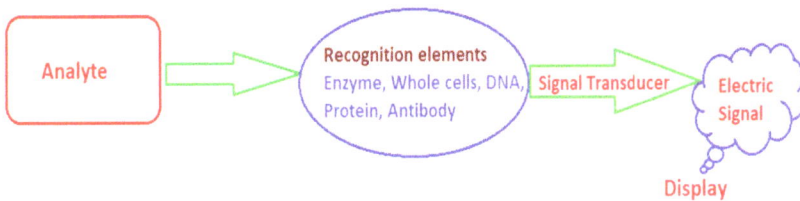

Fig. 1 The significant parts of the biosensor.

A typical case of a biosensor is a strategy that tests blood sugar, which applies glucose oxide to continually break down sugar in the blood. Thus, glucose is oxidized first and then two electrons are used to diminish flavin adenine dinucleotide (FAD) (one ingredient of the enzyme) until flavin Adenine Dinucleotide (FADH$_2$) shines [8]. At this stage, oxidation is carried out by the electrode (receiving two electrons from the electrode) in several steps where it produces a quantity of concentrated glucose, in this case, the electrical transformer and the enzyme are the biologically active elements [8].

The biosynthesis is performed to merge with the definitive analysis involved to provide an effect that can be measured by the converter. Significant selectivity for investigation between a matrix of other chemical or biological ingredients is a prerequisite for a biomarker. While the type of biomolecule that is extensively applied can differ, biosensors can be categorized corresponding to frequent types of bio-receptor interactions that incorporate: antibody/antigen [9], enzymes/ligaments, nucleic acids/DNA, structures/biological cells, or biomimetic materials [10].

2. General characteristics of the biosensor

The biosensor possesses two distinctive types of ingredients: (a) biological, for example, an enzyme, antibody, and (b) physical, for example, a transducer, amplifier [11]. The biological component of the biosensor offers two critical functions: (a) it recognizes the analysis and (b) it interacts with it in a way that generates some physical changes that can be revealed by the converter [11,12]. These properties of the biological component convey respectively the biosensor, the sensitivity and the ingenuity to detect and measure analysis. The biological component is appropriately equipped on the strength transformer. In general, the amended insensibility of enzymes improves their durability [12]. As a result, many enzyme-constrained strategies can be managed further than 10,000 times for several months.

The biological ingredient specifically interacts with the analyte that provides physical exchange near the surface of the transducer. This physical difference may be [13]:

1. The heat liberated or absorbed by the reaction (caloric biological sensors)

2. Creation of electrical potential because of changing electron configuration (potentiometric sensors).

3. The amperometric biological sensor, which is a movement of electrons thanks to reaction and reduction reactions.

4. The optical sensor refers to light produced or absorbed during the reaction.

5. Sound wave sensors, where the mass of the biological part changes because of the interaction.

The transformer distinguishes, measures and converts these transformations into the electrical signals. These signals are amplified by a subwoofer before being inserted into the microprocessor [14]. The signal is then processed, understood, and displayed in appropriate units. Consequently, the biosensors transform the flow of chemical signals into the flow of electrical information, which includes the subsequent steps: (a) The analysis disperses from the solution to the surface of the biosensors. (B) The preparation interacts concretely and effectively with the biological ingredient of the "biosensor". (C) This reaction alters the physical-chemical properties of the transformer surface. (D) This leads to a variation in the optical or electronic characteristics of the transformer surface. (E) The diversity is measured In optical/electronic properties, it is transformed into an electrical signal that is magnified, processed and displayed [14,15].

2.1 Biological calories

Several catalytic enzyme reactions are exothermic. The calorimeter sensors measure the change in the solution's temperature containing the analysis after the enzyme procedure and its interpretation in terms of the concentration of the analysis in the solution [16]. The analytical solution is passed through a small, packed column containing a fixed enzyme; the temperature of the solution is determined before the solution is inserted into the column just as the column is left using a separate thermostat [16,17]. The biggest disadvantage is maintaining the sample current temperature, approximately $\pm 0.01°C$. So, the sensitivity can be raised by applying two or more path enzymes in the biometric sensor to link multiple reactions to intensify heat yield. Alternatively, multifunctional enzymes can be used. An example of this is the use of glucose oxidase to determine glucose [12,18].

2.2   Potentiometric biological sensor

These sensors utilize ion-selective terminals to transform a biological reaction into an electronic signal [18,19].The terminals applied are the most prevalent glass electrodes in the pH adjust (for cations), the glass pH electrodes covered with a gas selective membrane [carbon dioxide(CO_2), ammonia (NH_3), or hydrogen sulfide (H_2S)]or solid-state electrodes [20]. Several reactions generate or adopt hydrogen ion (H^+) that are disclosed and calculated by the biosensor; in such processes, exceedingly weak stored solutions are managed. Gas sensing electrodes reveal and measure the amount of gas yielded [20]. An example of these electrodes is based on urease that arouses the following reactions

$$CO\,(NH_2)_2 + 2H_2O + H^+ \rightarrow 2NH_4^+ + HCO_3^-$$

This reaction can be estimated by pH responsiveness, pH sensor, NH_3 sensor or CO_2 electrode sensor. Biosensors can promptly be prepared by placing enzyme-coated membranes on the discriminating ion gates of particular transistors [21]. These vital sensors are very diminutive.

2.3   Sound waves

These use piezoelectric materials. Their surface is frequently covered with antibodies that attach to the reciprocal antigen present in the specimen solution. This strengthens the mass, which reduces the vibratory frequency. This variation is applied to demonstrate the number of antigens present in the sample solution [22, 23].

2.4   Biometric sensor

These electrodes work by providing a current when the voltage is referred between two electrodes. These electrodes work by providing a current when the voltage is connected between two electrodes, where, the intensity of the current commensurate to the concentricity of the substrate [24]. The simplest amperometry biological sensor utilizes the oxygen Clark pole that determines the oxygen (O_2) reductase performing in the sample solution (analyzer). These are the initial generation of biological sensors. These biosensors are exploited to measure oxidation and reduction feedbacks, an ordinary example is the dedication of the level of glucose using glucose oxidase [25].The main obstacle with allergens is their dependence on the dissolved O_2 concentration in the solution. This can be vanquished by using intermediaries; these molecules transport the electrons formed by the reaction directly to the electrode instead of reducing the dissolved O_2 in solution. These are also called second-generation biological sensors. However, current electrodes remove electrons instantaneously from the diminished enzymes without the benefit of mediators and are smeared with electrically conductive organic salts [26].

2.5   Optical sensor

These vital sensors measure stimulus and familiarity interactions. It measures the change in radiance or absorption caused by-products from catalytic reactions [27]. Alternatively, it measures the changes that have occurred in the intrinsic optical properties of the surface of the biosensor because it is loaded onto molecules as a dielectric (in the case of affinity reactions). A further promising biometric sensor that utilizes luciferase enzyme luminescence is applied to reveal microorganisms in food or clinical specimens [28]. The microorganisms are lysed to release adenosine triphosphate (ATP), which is used by luciferase in the existence of oxygen to yield light that is measured by biological sensing [28].

3.   Biosensors to detect toxins in the environment

Environmental surveillance is one of the great considerations for the integrity of living organisms from several toxic pollutants in the surrounding [29]. Various actions and scientific and social apprehensions have been argued to curb and determine the hazards of environmental deterioration. Accordingly, there is a requiring for promoting particular sensitive, expeditious, and discriminatory approaches that can reveal and evaluate the toxins for influential bioremediation procedure [29]. In this context, detached catalytic agents or vital procedures generating enzymes, as full cells or in the immobilized or latency status, can a provenance for disclosure, and deterioration or alteration of toxins to non-polluting compounds to improve the ecologically sound. The biological sensors are quintessential for the detection and appraisal of different toxins and pollutants in an authoritative, explicit, and oversensitive means [30]. The increment, in living criteria and greater consumer needs, has augmented deterioration of air purity, in the lack of necessary oxygen while conservation carbon dioxide and otherwise noxious gases. Besides, particulate matter of water accompanied with an assortment of synthetic substances and disposal of nonbiodegradable materials in the soil [31]. Synthetic chemicals like insecticides, cosmetics, personal, and household care products, and medicines, which are in handling global and are imperative for the contemporary community, are boosting toxins to the surrounding [32]. Humanitarian endeavors have arisen in the detection of contagion of water sources with organic micro-pollutants, like viruses and microorganisms which transfers to the living organisms. Such agents especially which conveyed by water cause many diseases and responsible for a death globally [32]. Biological sensors are strategies that use any biological technology to detect an analyte, whether toxins, pollutants or otherwise. The biosensor is unquestionably a sovereign component but is considered as an ingredient of a planned instrumentation [26].

Homogeneously, the sign qualities and advantages through the progression of an enzyme-based biosensor build on the efficiency and sensitivity during the assessment, precision, and authenticity to assessing, frequency and prolonged-term establishment, intensity, safety and flexibility, and acceptability by users [33]. Biological catalysts (enzymes) can distinguish the existence of definite analyzes by testing either the utilization or manufacture of definite compounds such as CO_2, NH_3, H_2O_2 or O_2 therefore, the probes determine the different toxins and tie in their presence into the substrates [33].

Newly promoted an expeditious and remarkably susceptible detection procedure for endotoxin. These studies point out that the initial revealing of endotoxin facilitates the administration of an antibiotic remedy and restricts the onslaught of sepsis. Sepsis is emanating from the presence of microbes and other venoms in the blood. This infection is the broadest doctrine of fatality in intensive care units. Some studies marked out that the quick disclosure of sepsis considerably aids in retrieving patient endurance [34].

The fundamental parts of a biosensor are the bio-realization ingredient, then the associate which extends the immobilization of the bioreceptor, finally the separate part which is a transducer. Regarding legitimate-time air pollution surveys, biomedical investigations have encouraged increased attention related to improvements in gene management that increase the likelihood of actual use of immovable microorganisms in illuminated lightings in the presence of specialized analyses [35]. Different genetic methods also allow pathogens to be programmed to respond in a similar manner to a specific class of compounds. When pathogens (whole cell sensors) are exploited, they can reveal toxins and yield accurate information about the biological effects associated with the toxicant [36].

The usefulness of living organisms exploring toxins is a valuable and trustworthy approach to evaluate the toxicity of unidentified specimens. Scientists note that the technologies that incorporate vital microorganisms are responsible for the delicate distinction between unknown toxins in water, soil, and shipwrecks. A wholesome cell biosensor is the usage of a Pseudomonas strain as an understanding matter in a biosensor device inside the flow over cell measuring sample injector to distinguish low concentrations of naphthalene [37,38]. The smaller curb of naphthalene detection through *Pseudomonas fluorescens* (*P. fluorescens*) strain (HK44) is below 0.1 mg/L (at 0.16 µm) as shown by Environmental Protection Agency (EPA).

It is normal for environmental toxins within the role or gas being investigated to be unusually unsuccessful. Non-selective adsorption regulates the results obtained. In these sorts of systems, it is necessary to reduce this non-selective adsorption [39]. This non-selective adsorption will not be in some experiments a huge component of the adsorbent material and that this coupling rate, which is a transitional rate, may depend on the availability of the surface [40]. Different microbial biosensors for the different toxic waste in the environment as shown in (Table 1) [41-56].

One of the most common methods used to detect mycotoxins is the liquid phase chromatography, which is a sensitive but more expensive method that consumes a long time in extraction and washing processes, as well as the thin layer chromatography and the gas phase and Eliza dishes, but the need remains urgent. Other methods are accurate, simple and less expensive than these methods for the rapid control of mycotoxins [57,58]. Among the biological methods used is a set of rapid biological tests that are easy to implement and can be performed without the need for sophisticated and costly devices, and these tests reveal the strains producing mycotoxins without determining the quality and quantity of mycotoxins, which can be determined later by performing other complementary tests [59]. Also, these tests are considered an important alternative in the absence of the possibility of chromatographic analysis and the absence of standard toxic substance [60]. As well as, the diversity of the organisms used in such tests as the use of a group of bacterial species of the genus *Bacillus* to detect toxins secreted from fungi such as *Fusarium equisetum, Alternariaalternata* that affect some types of fruits, as it was found from these tests that the toxins in these fungi prevented the development and growth of bacteria [61]. The presence of the genus *Fusarium* may be evidence of the presence of toxins such as *Zearalenone*, which causes hormonal disorders, enlarged genital system, atrophy of ovaries and testes in animals [62]. Many species of this species also produce another group of toxins known as *Trichothecenes*, which are found in both food and feed, and their ingestion causes gastrointestinal bleeding and vomiting, and direct contact leads to skin infections [63].

Table 1 Considerable biosensors advanced for the disclosure of particular pernicious composites [41-56].

Analytes	Biosensing elements	Transducers	Samples	References
I- Toxic metals *Nickel, zinc, and cadmium*	on a plasmid	Optical(luminometer)	Soil	[41]
Cadmium	DNA *Pseudomonas fluorescens* 10586s pUCD607with thelux insertion	Electrochemical	Standard solutions	[42]
Cadmium, arsenic and mercury	Urease enzyme	Electrochemical	Standard solutions	[43]
Aluminum, zinc, nickel, copper, lead, cadmium, andiron	*Chlorella vulgaris* strain CCAP 211/12	Electrochemical	Urban waters	[44]
Copper, Zinc, and cobalt	*Pseudomonas* sp. B4251, *Bacillus cereus* B4368 and E. coli 1257	Electrochemical	Water	[45]
Mercury and lead ions	DNA	Optical	H_2O	[46]
Cadmium, copper,and lead	Sol-gel-immobilized urease	Electrochemical	Syntheticceffluents	[47]
Organophosphates,urea, and ethanol	*Flavobacteium sp., Bacillus sp.,* and *S. ellipsoideus*	Potentiometric	Effluents andlaboratory samples	[48-50]
II- Phenolics				
phenol/chloro-phenol, catechol/phenol, cresol/chloro-cresol and phenol/cresol	Laccase and tyrosinase	Amperometry	Waste water (WW)	[51]
m-Cresol or catechol	DNA	Amperometry	WW	[44]
Phenol	Mushroom tissue (tyrosinase)	Amperometry	WW	[52]

III- Pesticides	Peroxidase	Potentiometric	Soil and waste	[53,54]
Simazina	(Biocatalytic)	Amperometry	water	
Parathion	Parathion	Optical	Soil	
Paraoxon	hydrolase	Amperometry	Soil and waste	
Carbaril	(Biocatalytic)		water	
	Alkaline		Soil and waste	
	phosphatase		water	
	Acetyl			
	cholinesterase			
IV- Herbicides	Acetyl	Amperometry	Soil	[55,56]
2,4-Dichloro-phenoxy	cholinesterase			
acetic acid				
Diuron, paraquat		Bioluminescence	Soil	
	Cyanobacterial			

Yeast fungus was very sensitive to patulin toxin when used at a concentration of 10 micrograms [64]. About 19 plants from several types of the cruciferous family were examined to study the effect of Aflatoxin B1 on seed germination and seedling development, and it was found that the cress plant was most sensitive to poison until the concentration of 8 micrograms of agar[65]. *Euglena gracilis* was also used to detect toxic fungal isolates from samples of fruit, vegetables, and crops. Contamination of soybean seeds and kernels with storage fungi and the mycotoxins they produce is dangerous for human and animal health as they lead to animal death or may cause various cancers and disorders[66]. Therefore, the toxicity of fungal isolates of soybean seed and kernel was detected using some vital tests that do not require complicated devices and significant costs, by using the seeds of *Lepidium sativum* [67].

Recently a study revealed that detection procedures for objective proteins are in need; exclusively those that are sure, responsive, discriminatory, and without intervention from backdrop elements [68]. Immunoassays are extensively utilized in bioanalytical methods [68,69],show out that all immunoassays use antibodies as capture molecules for durable and definite connection to target antigens. Some studies disclose that immunoassay performances have assay precision, accessibility, low expense, and reagent stability [70].

Infection with different types of toxins raises great controversy, due to its great importance at the scientific level. Public disclosure of various types of the discovery of toxins and pollutants on vital surfaces will reach a great place, to eliminate the various types of syndromes that are raised by these toxins, along with other syndromes that may be indirectly involved in their creation [71]. There is a need to expand more sensitive sensors to detect of harmful toxins and pollutants faster before they develop into patterns

Materials Research Forum LLC
https://doi.org/10.21741/9781644901175-7

of difficult-to-control diseases. Removing pollutants and toxins from streams, or regular drinking sources is known to be time-consuming, and at a very high, cost [72]. The sooner this obstacle is perceived, the sooner it can be the more it is possible to keep those sources away from sources of pollution, and to obtain vital ways that are compatible with the recommended standards set by various legislative bodies, government agencies and others [72].

The performance of the biosensors is based on the transfer of electrons between the sensor electrode and the enzyme, as the performance of the sensors increases with the decrease in the distance between the active sites of the enzyme and the electrode surface [73]. Redox and enzyme enzymes have begun to escalate as typical components of biometric sensors, given that their ability to achieve the process of transferring electrons is complementary to their accuracy in directed bonding and the catalytic activity of the process [74]. Because lactate may be associated with serious medical conditions, its discovery and monitoring are of importance in the healthcare field. For example, high levels of lactate in the blood may be an indication of a lack of oxygen or the presence of other conditions that cause the production of this substance to increase or be removed from the blood inadequately [75]. Lactate is an acidic substance in nature, which helps break down glucose, and thus the energy production in the body continues. Therefore, it has become necessary to develop a biometric sensor that monitors the metabolic compound known as "lactate", the design of which revolves around the combination of a polymer transporting electrons and the enzyme "lactate oxidase", an enzyme that specifically stimulates the oxidation of lactate [76]. The researchers paired the lactate oxidase enzyme with the organic electrochemical transistor polymer. This polymer transporting electrons simultaneously acts as an active transformer and a powerful amplifier of signals, as it can receive electrons from the enzymatic reaction and then undergo several reductive reactions through several active oxidation and reduction sites [76]. Likewise, this polymer carries hydrophilic side chains that facilitate the inter-molecular interactions with the lactate oxidase, bringing the enzyme closer to the energy-transporting substance. This stimulates electrical contact, thus increasing the polymer's sensitivity to lactate. As well as, these interactions between the polymer and the enzyme avoid modifications to the surface of the electrode and preclude the use of media, which "simplifies the manufacture of the sensor" [76]. Electrochemical biosensors bind enzymes to the electrode through a damping system to facilitate noticeable selectivity and selectivity with limited feedback rate analysis [77].

4. Future directions

The field of biosensors to control toxins has made a tremendous development. Advances in deoxyribonucleic acid (DNA) and inheritances technologies will recommend the creation of more fully cellular organisms, which enhances the specificity and fluctuations of enzymatic transmitters [78]. In this regard, many bacterial biological sensors have been manufactured using whole genetically engineered cells to produce measurable signals when responding to stress factors. The production of environmentally beneficial compounds with little or no toxic toxicity, and buildings with computerized ventilation systems that can monitor the air, either indoor or outdoor, will solve the roots of the obstacle to toxins or pollutants carried through this air. This progress will support the creation of an ideal, sustainable ambient in the air that can enhance work efficiency and reduce negative health impacts [78]. A critical limitation of indoor pollutant monitoring procedures is their inefficiency in multiple analyte detection, indeed although maximum health effects are provoked using the vulnerability to a bunch of toxins. Consequently, the inevitable importance will be on the sensation of multi-array sensors that rely mainly on one of the delicate biofilms (enzymes or organisms in the whole cell) that release the detection of numbers of separate compounds [77,79].

Clean chemistry based on the green structures is the future of the invention of biological sensors in most developed countries. It uses elements and products that typically emit reduced amounts of chemical compounds and pollution at rates that are not harmful to human health [80].

Moreover, nanotechnology is an upcoming technology that continues to examine and evaluate various toxins, and to try to treat toxic substances as well, through the nanoscale [59, 80]. These nanoparticles absorb the pollutants or stimulate reactions rapidly due to both high surface area and energy, diminishing energy utilization throughout the analysis and consequently inhibiting the discharge of toxins [81,82]. Regarding the nanoscale, these elements can approach inaccessible regions and biotechnology can be produced on-site as well as facilitating the design of widely applicable biometric sensors [81]. Moreover, Carbon nanotubes can monitor the enzyme's activity. In a study, the installation of proteins on the carbon nanotubes wall was offered through a molecule association. Charged group-carrying proteins are based on the pH that may be a semi-conductive monowall electrostatic window that generates the possibility of building a nanoparticle-size or pH-sensitive protein [83]. Single-wall nanotubes were used as biological sensors and the controlled binding of the oxidation and reduction enzyme glucose oxidase to the side nanotube wall was also accomplished by binding to the molecule and causing electrical conduction changes. These sensors provide a new tool for biomolecular diagnostics. Through these techniques, the catalytic compounds were

Materials Research Forum LLC
https://doi.org/10.21741/9781644901175-7

successfully prepared by suspending the single-wall nanotubes and the alpha-chemotropic nanoparticles in the solution of polymethyl methacrylate in toluene [83]. Through the previous technique, it was observed that the alpha-chymotrypsin nanoparticles are higher than the alpha-chemotropic-graphite polymer. Besides, compounds containing single-wall nanotubes show higher enzymatic activity than preparations that do not contain these tubes [84].

The qualified nanomaterial sensors (QNS) are an outstanding generation that contributes particular disclosure, that have a size less than picomolar [82]. QNS consists of three components: a material (s) manufactured on the basis of nanoscale, the identification component that contributes to privacy, and a signal transfer technique that produces a means to transfer the presence of the analytical material [85]. Although these components are not always separate units within the sensor, each nano sensor may be characterized by the presence of these parts. Sensors can be made to identify individual or characteristic analyzes, called multiplex discovery [86].

Selectivity is an exceedingly essential aspect inside the construction of a prosperous biosensor. A varied pattern of understanding elements has been realized in nano sensory layouts such as antibodies [50], aptamer [87], enzymes , and functional proteins [88]. Similarly, Chen et al. [89] take advantage of the massive surface area of carbon nanotubes to design a biosensor for pesticides from organic phosphorus. The amino carbon nanotubes (CNT- NH_2) were dried on the surface of a standard glass electrode and were finally incubated with acetylcholine esterase (AChE). Using a differential pulse potentiometer, the value/ nanometer for that glass electrode was 0.08. Lately, Song et al. [90] mention the usefulness of the nanoscale component for advancing the creation of AChE biochemical sensors. A layer of low graphene oxide was used on a glass carbon electrode, which was observed by isolating the porous titanium dioxide (TiO_2) bound to chitosan, a biocompatible polymer. The coherence of the matrix was enhanced by electrophoresis of the second layer of chitosan to produce a medium-sized multi-layer nanostructure. The total detection time required is about 25 minutes, although the maximum disclosure for the sensor shown in Chen et al was further. There are many natural polymers such as chitosan with biocompatibility and biodegradability that are used to support living organisms or enzymes for use in biosensor applications, but they suffer from limited electrical conductivity that impedes biosensor activities [91]. Some scientists interested in the field of biosensors have developed nanoparticles with a core component of chitosan with super-branched polymeric chains and supported with metal nanorods such as gold and silver, and nanoparticles such as manganese dioxide and zinc oxide, to improve the biosensor's sensitivity to glucose molecules. These advanced hybrid nanomaterials can also be used as a substitute for commercially conducting inks [91].

Mercury detection processes were also developed by thiol by being bound to a variety of nanoparticles such as gold and silver, or quantum dots [91, 92]. Reaction based competition assays in which Hg(II) replaces a surface coating have likewise been interpreted in the literature [93,94]. Huang and Chang [95] created a sensor that transmits a fluorescence signal in the presence of mercury causing the displacement of rhodamine 6G from the surface of the nanoparticles. In the process of developing and updating mercury detection through thiol, 2.0 ppb of toxic mercury was detected in less than 10 minutes. In conjunction with medical treatments that are fundamentally based on biotechnologies, biotechnology techniques for toxins and environmental pollutants are nonetheless in their infancy and dealing with many challenges due to the intrinsic characteristics of environmental assessments [96]. The maximum disclosure and specificity of analyzes are key consequences for the promising improvement of biometric sensors in environmental analysis [96]. The paper-based device is also a promising biological system for mobile devices, which is shrinkable, low-cost, user-friendly, and meets the needs of immediate detection of environmental samples. The incorporation of nanomaterials into these biological sensors prepares new strategies to increase their analytical performance [97]. Although several biometric sensors have been developed for an extended set of environmental pollutants on a laboratory scale, a small number of biological sensors are commercially viable at the present time [98]. It is clear that more efforts are needed toward innovations that will play a reliable role in the development of automatic, continuous and actual sensors with high throughput analysis of environmental samples [98].

Conclusions

The biological toxins are noxious substances yielded by living cells or organisms that are intrusive at excessively low concentrations. Biosensing is observable from other physicochemical ways, such as mass spectrometry, that can significantly be remarkably impressionable and unequivocal in their own right. It could be used as a screening bio-techniques for the appraisal of the toxicity of a sample. In a massive majority of all referred to studies, antibodies were exploited to an extensive variety of targets because of their being specialized to the target analyte. With the noteworthy advances being established in science, subsequent sensors and biosensors are expected to demonstrate a noticeable enhancement in processing agility and competence, such that remarkably accurate portable sensing mechanisms and numerous analysis devices will be a critical provision for eventual biological platforms.

References

[1]   M. Badihi-Mossberg, V. Buchner, J. Rishpon, Electrochemical biosensors for pollutants in the environment, Electroanal. An Int. J. Devoted to Fundam. Pract. Asp. Electroanal. 19 (2007) 2015–2028. https://doi.org/10.1002/elan.200703946

[2]   S.J. Pearton, F. Ren, Y.L. Wang, B.H. Chu, K.H. Chen, C.Y. Chang, W. Lim, J. Lin, D.P. Norton, Recent advances in wide bandgap semiconductor biological and gas sensors, Prog. Mater. Sci. 55 (2010) 1–59. https://doi.org/10.1016/j.pmatsci.2009.08.003

[3]   F. Scheller, F. Schubert, Biosensors, Elsevier, 1991. https://doi.org/10.1016/ 0958-1669(91)90054-9

[4]   N.J. Ronkainen, H.B. Halsall, W.R. Heineman, Electrochemical biosensors, Chem. Soc. Rev. 39 (2010) 1747–1763. https://doi.org/10.1039/B714449K

[5]   S. Neethirajan, Recent advances in wearable sensors for animal health management, Sens. Bio-Sensing Res. 12 (2017) 15–29. https://doi.org/ 10.1016/j.sbsr.2016.11.004

[6]   M. Keusgen, Biosensors: new approaches in drug discovery, Naturwissenschaften. 89 (2002) 433–444. https://doi.org/10.1007/s00114-002-0358-3

[7]   A. Roda, P. Pasini, M. Guardigli, M. Baraldini, M. Musiani, M. Mirasoli, Bio-and chemiluminescence in bioanalysis, Fresenius. J. Anal. Chem. 366 (2000) 752–759. https://doi.org/10.1007/s002160051569

[8]   J.D. Newman, A.P.F. Turner, Home blood glucose biosensors: a commercial perspective, Biosens. Bioelectron. 20 (2005) 2435–2453. https://doi.org/ 10.1016/j.bios.2004.11.012

[9]   A. Juzgado, A. Soldà, A. Ostric, A. Criado, G. Valenti, S. Rapino, G. Conti, G. Fracasso, F. Paolucci, M. Prato, Highly sensitive electrochemiluminescence detection of a prostate cancer biomarker, J. Mater. Chem. B. 5 (2017) 6681–6687. https://doi.org/10.1039/C7TB01557G

[10]   G. Valenti, E. Rampazzo, E. Biavardi, E. Villani, G. Fracasso, M. Marcaccio, F. Bertani, D. Ramarli, E. Dalcanale, F. Paolucci, An electrochemiluminescence-supramolecular approach to sarcosine detection for early diagnosis of prostate cancer, Faraday Discuss. 185 (2015) 299–309. https://doi.org/10.1039/C5FD00096C

[11]   S.P. Mohanty, E. Kougianos, Biosensors: a tutorial review, Ieee Potentials. 25 (2006) 35–40. https://doi.org/10.1109/MP.2006.1649009

Materials Research Forum LLC
https://doi.org/10.21741/9781644901175-7

[12]   F.-G. Banica, Chemical sensors and biosensors: fundamentals and applications. John Wiley & Sons, 2012.

[13]   D.R. Thévenot, K. Toth, R.A. Durst, G.S. Wilson, Electrochemical biosensors: recommended definitions and classification, Anal. Lett. 34 (2001) 635–659. https://doi.org/10.1081/AL-100103209

[14]   P.D. Patel, (Bio) sensors for measurement of analytes implicated in food safety: a review, TrAC Trends Anal. Chem. 21 (2002) 96–115. https://doi.org/10.1016/S0165-9936(01)00136-4

[15]   Z. Farka, T. Juřík, D. Kovář, L. Trnková, P. Skládal, Nanoparticle-based immunochemical biosensors and assays: recent advances and challenges, Chem. Rev. 117 (2017) 9973–10042. https://doi.org/10.1021/acs.chemrev.7b00037

[16]   M.S. Thakur, K. V Ragavan, Biosensors in food processing, J. Food Sci. Technol. 50 (2013) 625–641. https://doi.org/10.1007/s13197-012-0783-z

[17]   S.R. Mikkelsen, E. Cortón, Bioanalytical chemistry, John Wiley & Sons, 2016. https://doi.org/10.1021/np058236+

[18]   C. Chen, X.-L. Zhao, Z.-H. Li, Z.-G. Zhu, S.-H. Qian, A.J. Flewitt, Current and emerging technology for continuous glucose monitoring, Sensors. 17 (2017) 182. https://doi.org/10.3390/s17010182

[19]   N. Gupta, S. Sharma, I.A. Mir, D. Kumar, Advances in sensors based on conducting polymers, (2006). http://hdl.handle.net/123456789/4862

[20]   W.E. Morf, The principles of ion-selective electrodes and of membrane transport, Elsevier, 2012.

[21]   B.R. Eggins, Biosensors: an introduction, Springer-Verlag, 2013. https://doi.org/10.1007/978-3-663-05664-5 .

[22]   F.G. Barth, J.A.C. Humphrey, T.W. Secomb, Sensors and sensing in biology and engineering, Springer Science & Business Media, 2003.

[23]   H. Mansy, R. Sandler, Sensors and sensor assemblies for monitoring biological sounds and electric potentials, (2004). https://patents.google.com/patent/US20040032957A1/en

[24]   S. Ikeda, T. Yoshioka, S. Nankai, H. Tsutsumi, H. Baba, Y. Tokuno, S. Miyazaki, Biosensor, and a method and a device for quantifying a substrate in a sample liquid using the same, (1997). https://patents.google. com/patent/US5582697A/en

[25] M.M. Rahman, A. Umar, K. Sawada, Development of amperometric glucose biosensor based on glucose oxidase co-immobilized with multi-walled carbon nanotubes at low potential, Sensors Actuators B Chem. 137 (2009) 327–333. https://doi.org/10.1016/j.snb.2008.10.060

[26] S. Vigneshvar, C.C. Sudhakumari, B. Senthilkumaran, H. Prakash, Recent advances in biosensor technology for potential applications–an overview, Front. Bioeng. Biotechnol. 4 (2016) 11. https://doi.org/10.3389/fbioe.2016.00011

[27] J.S. Schultz, Optical sensor of plasma constituents, (1982).

[28] P. Banerjee, A.K. Bhunia, Mammalian cell-based biosensors for pathogens and toxins, Trends Biotechnol. 27 (2009) 179–188. https://doi.org/10.1016/j.tibtech.2008.11.006

[29] J.M. Hellawell, Biological indicators of freshwater pollution and environmental management, Springer Science & Business Media, 2012.

[30] H. Sharma, M. Agarwal, M. Goswami, A. Sharma, S.K. Roy, R. Rai, M.S. Murugan, Biosensors: tool for food borne pathogen detection, Vet. World. 6 (2013) 968. https://doi.org/10.14202/vetworld.2013.968-973

[31] F.R. Spellman, The science of environmental pollution, Crc Press, 2017. https://doi.org/10.1201/9781315226149

[32] D. Dhaniram, Chemicals of emerging concern in household products: a case study on the disposal of cosmetics in the United Kingdom, (2011). https://doi.org/10.25560/9281.

[33] S. Hassani, S. Momtaz, F. Vakhshiteh, A.S. Maghsoudi, M.R. Ganjali, P. Norouzi, M. Abdollahi, Biosensors and their applications in detection of organophosphorus pesticides in the environment, Arch. Toxicol. 91 (2017) 109–130. https://doi.org/10.1007/s00204-016-1875-8

[34] I. Karlsson, Cytokines as diagnostic biomarkers in canine pyometra and sepsis, 2015. https://doi.org/10.1016/j.theriogenology.2015.02.008

[35] R. Khot, V. Chitre, Survey on air pollution monitoring systems, in: 2017 Int. Conf. Innov. Information, Embed. Commun. Syst., IEEE, 2017: pp. 1–4. https://doi.org/10.1109/ICIIECS.2017.8275846

[36] I. Ahmed, Z. Akram, M.H. Bule, H. Iqbal, Advancements and potential applications of microfluidic approaches—a review, Chemosensors. 6 (2018) 46. https://doi.org/10.3390/chemosensors6040046

[37] M. Gronow, Biosensors, Trends Biochem. Sci. 9 (1984) 336–340.
https://doi.org/10.1016/0968-0004(84)90055-0

[38] D.J. Pike, N. Kapur, P.A. Millner, D.I. Stewart, Flow cell design for effective
biosensing, Sensors. 13 (2013) 58–70. https://doi.org/10.3390/s130100058

[39] A. Kot, J. Namiesńik, The role of speciation in analytical chemistry, TrAC Trends
Anal. Chem. 19 (2000) 69–79. https://doi.org/10.1016/S0165-9936(99)00195-8

[40] E. Valdman, I.G.R. Gutz, Bioluminescent sensor for naphthalene in air: Cell
immobilization and evaluation with a dynamic standard atmosphere generator, Sensors
Actuators B Chem. 133 (2008) 656–663. https://doi.org/10.1016/ j.snb.2008.03.031

41] S.P. McGrath, B. Knight, K. Killham, S. Preston, G.I. Paton, Assessment of the
toxicity of metals in soils amended with sewage sludge using a chemical speciation
technique and a lux-based biosensor, Environ. Toxicol. Chem. An Int. J. 18 (1999)
659–663.

[42] E.L.S. Wong, E. Chow, J.J. Gooding, The electrochemical detection of cadmium
using surface-immobilized DNA, Electrochem. Commun. 9 (2007) 845–849.
https://doi.org/10.1002/etc.5620180411

[43] P. Pal, D. Bhattacharyay, A. Mukhopadhyay, P. Sarkar, The detection of mercury,
cadium, and arsenic by the deactivation of urease on rhodinized carbon, Environ. Eng.
Sci. 26 (2009) 25–32. https://doi.org/10.1016/j.elecom.2006.11.018

[44] D. Claude, G. Houssemeddine, B. Andriy, C. Jean-Marc, Whole cell algal
biosensors for urban waters monitoring, in: Novatech, 2007: pp. 1507–1514.
https://doi.org/10.1089/ees.2007.0148

[45] T.G. Gruzina, A.M. Zadorozhnyaya, G.A. Gutnik, V. V Vember, Z.R. Ulberg, N.I.
Kanyuk, N.F. Starodub, A bacterial multisensor for determination of the contents of
heavy metals in water, J. Water Chem. Technol. 29 (2007) 50–53.
https://doi.org/10.1136/jech.2006.049205

[46] M.R. Knecht, M. Sethi, Bio-inspired colorimetric detection of Hg 2+ and Pb 2+
heavy metal ions using Au nanoparticles, Anal. Bioanal. Chem. 394 (2009) 33–46.

[47] R. Ilangovan, D. Daniel, A. Krastanov, C. Zachariah, R. Elizabeth, Enzyme based
biosensor for heavy metal ions determination, Biotechnol. Biotechnol. Equip. 20
(2006) 184–189. https://doi.org/10.1080/13102818.2006.10817330

[48] S. Gäberlein, F. Spener, C. Zaborosch, Microbial and cytoplasmic membrane-
based potentiometric biosensors for direct determination of organophosphorus

insecticides, Appl. Microbiol. Biotechnol. 54 (2000) 652–658.
https://doi.org/10.1007/s002530000437

[49]   N. Verma, M. Singh, A disposable microbial based biosensor for quality control in milk, Biosens. Bioelectron. 18 (2003) 1219–1224. https://doi.org/10.1016/S0956-5663(03)00085-X

[50]   L. Rotariu, C. Bala, V. Magearu, New potentiometric microbial biosensor for ethanol determination in alcoholic beverages, Anal. Chim. Acta. 513 (2004) 119–123. https://doi.org/10.1016/j.aca.2003.12.048

[51]   H.B. Yildiz, J. Castillo, D.A. Guschin, L. Toppare, W. Schuhmann, Phenol biosensor based on electrochemically controlled integration of tyrosinase in a redox polymer, Microchim. Acta. 159 (2007) 27–34. https://doi.org/10.1007/s00604-007-0768-1

[52]   F. Karim, A.N.M. Fakhruddin, Recent advances in the development of biosensor for phenol: a review, Rev. Environ. Sci. Bio/Technology. 11 (2012) 261–274. https://doi.org/10.1007/s11157-012-9268-9

[53]   S. Rodriguez-Mozaz, M.J.L. de Alda, D. Barceló, Biosensors as useful tools for environmental analysis and monitoring, Anal. Bioanal. Chem. 386 (2006) 1025–1041. https://doi.org/10.1007/s00216-006-0574-3

[54]   G.A.E. Mostafa, Electrochemical biosensors for the detection of pesticides, Open Electrochem. J. 2 (2010). https://doi.org/10.2174/1876505X01002010022]

[55]   A. Sassolas, L.J. Blum, B.D. Leca-Bouvier, Immobilization strategies to develop enzymatic biosensors, Biotechnol. Adv. 30 (2012) 489–511. https://doi.org/10.1016/j.biotechadv.2011.09.003

[56]   A. Hayat, J.L. Marty, Aptamer based electrochemical sensors for emerging environmental pollutants, Front. Chem. 2 (2014) 41. https://doi.org/10.3389/fchem.2014.00041

[57]   A.C. Patel, S. Li, J.-M. Yuan, Y. Wei, In situ encapsulation of horseradish peroxidase in electrospun porous silica fibers for potential biosensor applications, Nano Lett. 6 (2006) 1042–1046. https://doi.org/10.1021/nl0604560

[58]   A.P. Wacoo, D. Wendiro, P.C. Vuzi, J.F. Hawumba, Methods for detection of aflatoxins in agricultural food crops, J. Appl. Chem. 2014 (2014). https://doi.org/10.1155/2014/706291

[59]   A.F. Sahab, A.I. Waly, M.M. Sabbour, L.S. Nawar, Synthesis, antifungal and insecticidal potential of chitosan (CS)-g-poly (acrylic acid)(PAA) nanoparticles

against some seed borne fungi and insects of soybean, Int. J. ChemTech Res. 8 (2015) 589–598.

[60]   J. Yang, J. Li, Y. Jiang, X. Duan, H. Qu, B. Yang, F. Chen, D. Sivakumar, Natural occurrence, analysis, and prevention of mycotoxins in fruits and their processed products, Crit. Rev. Food Sci. Nutr. 54 (2014) 64–83. https://doi.org/10.1080/10408398.2011.569860

[61]   C. Kosawang, M. Karlsson, B. Jensen, H. Vélëz, P.H. Rasmussen, D.B. Collinge, D.F. Jensen, Detoxification of the Fusarium mycotoxin zearalenone is an important trait of Clonostachys rosea in biocontrol of Fusarium foot rot of barley., in: Work. Gr. "Biological Control Fungal Bac Terial Plant Pathog. Proc. Meet. Reims, Fr. 24–2 7 June 2012, 2013: pp. 133–136.

[62]   G.P. Munkvold, Fusarium species and their associated mycotoxins, in: Mycotoxigenic Fungi, Springer, 2017: pp. 51–106. https://doi.org/10.1007/978-1-4939-6707-0_4

[63]   W.M. Haschek, J.C. Haliburton, Fusarium moniliforme and zearalenone toxicoses in domestic animals: A review, in: Diagnosis of Mycotoxicoses, Springer, 1986: pp. 213–235. https://doi.org/10.1007/978-94-009-4235-6_20

[64]   I.E. Yates, J.K. Porter, Bacterial bioluminescence as a bioassay for mycotoxins., Appl. Environ. Microbiol. 44 (1982) 1072–1075.

[65]   R. Zhu, K. Feussner, T. Wu, F. Yan, P. Karlovsky, X. Zheng, Detoxification of mycotoxin patulin by the yeast Rhodosporidium paludigenum, Food Chem. 179 (2015) 1–5. https://doi.org/10.1016/j.foodchem.2015.01.066

[66]   M.A. El-Shafie, Method and system for processing a biomass for producing biofuels and other products, (2019). https://patents.google.com/patent/US20120122164A1/en

[67]   N.K. Rao, J. Hanson, M.E. Dulloo, K. Ghosh, A. Nowell, Manual of seed handling in genebanks, Bioversity International, 2006. https://doi.org/10.1017/S0014479707005741

[68]   A.J. Haes, A. Terray, G.E. Collins, Bead-assisted displacement immunoassay for staphylococcal enterotoxin B on a microchip, Anal. Chem. 78 (2006) 8412–8420. https://doi.org/10.1021/ac061057s

[69]   E.P. Diamandis, T.K. Christopoulos, Immunoassay, Academic Press, 1996. https://doi.org/10.1038/npg.els.0001135

[70]   A.E. Herr, D.J. Throckmorton, A.A. Davenport, A.K. Singh, On-chip native gel electrophoresis-based immunoassays for tetanus antibody and toxin, Anal. Chem. 77 (2005) 585–590. https://doi.org/10.1021/ac0489768

[71]   J.W. Grate, C.J. Bruckner-Lea, D.A. Holman, Flow-controlled magnetic particle manipulation, (2011). https://patents.google.com/patent/US7892856B2/en

[72]   O. Zielinski, J.A. Busch, A.D. Cembella, K.L. Daly, J. Engelbrektsson, A.K. Hannides, H. Schmidt, Detecting marine hazardous substances and organisms: sensors for pollutants, toxins, and pathogens, Ocean Sci. 5 (2009) 329. https://doi.org/10.5194/os-5-329-2009

[73]   P. Das, M. Das, S.R. Chinnadayyala, I.M. Singha, P. Goswami, Recent advances on developing 3rd generation enzyme electrode for biosensor applications, Biosens. Bioelectron. 79 (2016) 386–397. https://doi.org/10.1016/j.bios.2015.12.055

[74]   P.J. Delves, S.J. Martin, D.R. Burton, I.M. Roitt, Roitt's essential immunology, John Wiley & Sons, 2017.

[75]   G.A. Brooks, The science and translation of lactate shuttle theory, Cell Metab. 27 (2018) 757–785. https://doi.org/10.1016/j.cmet.2018.03.008

[76]   A.M. Pappa, D. Ohayon, A. Giovannitti, I.P. Maria, A. Savva, I. Uguz, J. Rivnay, I. McCulloch, R.M. Owens, S. Inal, Direct metabolite detection with an n-type accumulation mode organic electrochemical transistor, Sci. Adv. 4 (2018) eaat0911. https://doi.org/10.1126/sciadv.aat0911

[77]   J. Janata, Principles of chemical sensors, Springer Science & Business Media, 2010. https://doi.org/10.1007/978-0-387-69931-8 1

[78]   S. Kintzios, P. Banerjee, Mammalian cell-based sensors for high throughput screening for detecting chemical residues, pathogens, and toxins in food, in: High Throughput Screen. Food Saf. Assess., Elsevier, 2015: pp. 123–146. https://doi.org/10.1016/B978-0-85709-801-6.00005-8

[79]   P. Biswas, C.-Y. Wu, Nanoparticles and the environment, J. Air Waste Manage. Assoc. 55 (2005) 708–746. https://doi.org/10.1080/10473289.2005.10464656

[80]   E.S. Beach, Z. Cui, P.T. Anastas, Green chemistry: A design framework for sustainability, Energy Environ. Sci. 2 (2009) 1038–1049. https://doi.org/10.1039/B904997P

[81]   H. Chen, R. Yada, Nanotechnologies in agriculture: new tools for sustainable development, Trends Food Sci. Technol. 22 (2011) 585–594. https://doi.org/10.1016/j.tifs.2011.09.004

[82] R.J. Miller, S. Bennett, A.A. Keller, S. Pease, H.S. Lenihan, TiO2 nanoparticles are phototoxic to marine phytoplankton, PLoS One. 7 (2012). https://doi.org/10.1371/journal.pone.0030321

[83] Y. Zhou, Y. Fang, R.P. Ramasamy, Non-covalent functionalization of carbon nanotubes for electrochemical biosensor development, Sensors (Switzerland). 19 (2019). https://doi.org/10.3390/s19020392.

[84] A.A. Chaudhari, S. deb Nath, K. Kate, V. Dennis, S.R. Singh, D.R. Owen, C. Palazzo, R.D. Arnold, M.E. Miller, S.R. Pillai, A novel covalent approach to bio-conjugate silver coated single walled carbon nanotubes with antimicrobial peptide, J. Nanobiotechnology. 14 (2016) 58. https://doi.org/10.1186/s12951-016-0211-z

[85] P. Makaram, D. Owens, J. Aceros, Trends in nanomaterial-based non-invasive diabetes sensing technologies, Diagnostics. 4 (2014) 27–46. https://doi.org/10.3390/diagnostics4020027

[86] L.B. Sagle, L.K. Ruvuna, J.A. Ruemmele, R.P. Van Duyne, Advances in localized surface plasmon resonance spectroscopy biosensing, Nanomedicine. 6 (2011) 1447–1462. https://doi.org/10.2217/nnm.11.117

[87] P. Shukla, V. Nigam, R. Gupta, A. Singh, R.C. Kuhad, Sustainable enzyme technology for environment: biosensors for monitoring of pollutants and toxic compounds, in: Biotechnol. Environ. Manag. Resour. Recover., Springer, 2013: pp. 69–76. https://doi.org/10.1007/978-81-322-0876-1_4

[88] N. Li, C.-M. Ho, Patterning functional proteins with high selectivity for biosensor applications, JALA J. Assoc. Lab. Autom. 13 (2008) 237–242. https://doi.org/10.1016/j.jala.2008.04.001

[89] Y. Chen, S. Zhou, L. Li, J. Zhu, Nanomaterials-based sensitive electrochemiluminescence biosensing, Nano Today. 12 (2017) 98–115. https://doi.org/10.1016/j.nantod.2016.12.013

[90] Y. Song, Y. Luo, C. Zhu, H. Li, D. Du, Y. Lin, Recent advances in electrochemical biosensors based on graphene two-dimensional nanomaterials, Biosens. Bioelectron. 76 (2016) 195–212. https://doi.org/10.1016/j.bios.2015.07.002

[91] E. Vunain, A.K. Mishra, B.B. Mamba, Fundamentals of chitosan for biomedical applications, in: J.A. Jennings, J.D. Bumgardner (Eds.) Chitosan based biomater. Vol. 1, Elsevier, 2017: pp. 3–30. https://doi.org/10.1016/B978-0-08-100230-8.00001-7

[92]   J. Du, L. Jiang, Q. Shao, X. Liu, R.S. Marks, J. Ma, X. Chen, Colorimetric detection of mercury ions based on plasmonic nanoparticles, Small. 9 (2013) 1467–1481. https://doi.org/10.1002/smll.201200811

[93]   P. Valentini, P.P. Pompa, Gold nanoparticles for naked-eye DNA detection: smart designs for sensitive assays, RSC Adv. 3 (2013) 19181–19190. https://doi.org/10.1039/C3RA43729A

[94]   M. Sabela, S. Balme, M. Bechelany, J. Janot, K. Bisetty, A review of gold and silver nanoparticle-based colorimetric sensing assays, Adv. Eng. Mater. 19 (2017) 1700270. https://doi.org/10.1002/adem.201700270

[95]   C.C. Huang, H.-T. Chang, Selective gold-nanoparticle-based "turn-on" fluorescent sensors for detection of mercury (II) in aqueous solution, Anal. Chem. 78 (2006) 8332–8338. https://doi.org/10.1021/ac061487i

[96]   J. Yao, M. Yang, Y. Duan, Chemistry, biology, and medicine of fluorescent nanomaterials and related systems: new insights into biosensing, bioimaging, genomics, diagnostics, and therapy, Chem. Rev. 114 (2014) 6130–6178. https://doi.org/10.1021/cr200359p

[97]   X. Wang, X. Lu, J. Chen, Development of biosensor technologies for analysis of environmental contaminants, Trends Environ. Anal. Chem. 2 (2014) 25–32. https://doi.org/10.1016/j.teac.2014.04.001

[98]   V.K. Nigam, P. Shukla, Enzyme based biosensors for detection of environmental pollutants-A review, J. Microbiol. Biotechnol. 25 (2015) 1773–1781. https://doi.org/10.4014/jmb.1504.04010

Toxic Gas Sensors and Biosensors
Materials Research Foundations **92** (2021) 220-244

Materials Research Forum LLC
https://doi.org/10.21741/9781644901175-8

Chapter 8

Colorimetric and Fluorometric Sensor Arrays

A. Kantürk Figen[1]*, Y. Basaran Elalmis[2], B. Coşkuner Filiz[3]

[1]Yildiz Technical University, Chemical Engineering Department, Istanbul, Turkey

[2]Yildiz Technical University, Bioengineering Department, Istanbul, Turkey

[3]Yildiz Technical University, Science and Technology Application and Research Center, Istanbul, Turkey

*akanturk@yildiz.edu.tr

Abstract

The design and construction of colorimetric/fluorometric sensor arrays with high selectivity and sensitivity has been of considerable attention as an emerging technology for mobile chemical detection. Nowadays, many approaches have been made to design sensors, fabricate arrays and generalize their usage areas especially in daily life applications and industrial sectors. This chapter introduces the fundamentals, fabrication methods, and applications of colorimetric and fluorometric sensor arrays. The readers will find detailed information about mechanism of chemical sensing and optical sensor arrays; solid state sensor fabrication methodologies; and specific applications environmental, pharmaceutical, medical, and food sectors.

Keywords

Sensor, Colorimetric, Fluorometric, Arrays, Fabrication

Contents

Materials Research Forum LLC
https://doi.org/10.21741/9781644901175-8

1.   Introduction

Sensor systems basically consist of a sensing element by which the target analyte is detected, and a transducer to convert this detection to a readable signal [1]. Sensors can be divided into three types, which are physical, chemical and biosensors. Physical sensors are used for the measurement of a physical quantity such as distance, temperature and mass. Chemical sensors are used for the measurement of chemical materials by chemical/physical responses. Biosensors are also used for the measurement of chemical materials; however, they utilize biological sensing elements in these measurements. The specific chemicals measured by chemical and biosensors are usually referred to as substrate or analyte, which is a more general term. Substrate or analyte may be a biological or chemical material [2].

Chemical sensors have turned into important chemical analysis and detection tools during recent years. They were designed to be highly selective and thus efficient for many applications. However, it is not practical to design specific receptors for each analyte present in a multi-component mixture. On the other hand, array-based sensors involve many sensing elements thereby respond to various analytes via cross-reactivity among the analyte and array. Sensing elements present in the array do not need to be perfectly specific or selective for target substrates, some sensing elements respond at a different degree to the other substrates as well as the target substrate. Various substrates can be differentiated by means of distinctness of the overall response obtained from the sensor array.

Photochemical colorimetric method is frankly the most favorable and effective method among other detection methods such as optics, electrochemistry, and chromatography. Optical colorimetric/fluorometric sensor array has been extensively applied due to its properties such as low cost, fast response, simple operation, and multi-information [3]. Solid-state colorimetric sensors can bring some advantages on real time qualitative/semi-quantitative detection and on-site without complicated analytical instruments [4].

Colorimetric detection of a specific analyte is based on changing of absorption band of the material which results in color change. This analytical measurement method is easy to apply with simplified sensing apparatus that does not require complex and expensive instruments. Beside this, color change during detection can be affected by experimental conditions and the concentration of the target analyte should be measured by more accurate systems. So it requires absorbance measurements using a UV-vis spectrometer after the color change [5, 6]. Two main requirements should be taken into consideration for the fabrication of colorimetric sensor arrays. First one is associated with chemically-responsive pigment that must strongly interact with analytes and second one is this interaction should be combined with intense chromospheres [7]. Luminescence is spontaneous of light emission by a substance not resulting from heat depending on the nature of the excited state and emission rates and luminescence is divided into two categories, namely fluorescence and phosphorescence. In optical sensor application, sensor is hit by the photon and molecules release a fluorescence emission [8-9]. In this method, the intensity can be measure directly and reference is not required [10].

2.   Chemical sensing and optical sensors

2.1   Chemical sensing

Chemical sensors are devices which respond selectively to a specific analyte of interest via physical or chemical interactions, and are useful in the determination of the analyte qualitatively or quantitatively. Receptor and transducer are two main parts of the sensors. Receptor is the part which reacts or unites with the analyte and change its physical or chemical property, and transducer is the part which transforms and amplifies these changes into signal outputs relevant to the analyte properties. Signal output may be electrochemical, optical, etc. [3].

Traditional chemical sensing method originate from the "lock-and-key" model of molecular recognition and depends on the detection of target analytes using high selectivity receptors. This traditional sensing technique ideally includes a fully selective or specific sensor for a single target analyte, hence no other species is identified [11].

The "lock-and-key" model concept, which is substantially old, was first postulated by Emil Fischer in 1894 [12]. Specific enzyme catalyzed reactions can be described using this "lock-and-key" analogy. Enzymes which are highly specific biomolecules, are able to identify and bind to their substrate specifically. Here, the enzyme is represented with lock and substrate is represented with key. Enzymatic activity is initiated with the binding of substrate (key) to the active site (keyhole) of the enzyme (lock). It was hypothesized by Fischer that dual interaction between the enzyme and substrate was the driving force of the enzyme-substrate binding. Intermolecular interactions such as hydrogen bonding, hydrophobic and electrostatic interactions present in enzyme structures are generally used in sensing area due to their similarity to molecular recognition interactions. Molecular recognition (lock-and-key) model inspired the synthetic sensor design for years.

In lock-and-key model certain targets must be considered in the synthetic receptor design. For example, boronic acid moieties are usually integrated into the synthetic receptor structures as diol or sugar targeting receptor units and hydrophobic units are integrated as the binding sides for aliphatic chains.

Although specific sensing methods- utilizing single target- are highly advantageous and advanced, these sensing methods may not be feasible for the detection of more than two analytes simultaneously. To cope with the challenges present, molecular recognition in biological olfactory system has become the main focus of researchers now [13].

Search for novel chemical detection techniques resulted in development of several helpful sensors as substitute for electronic noses based on electrical responses. Optical sensors are particularly considerable among these.

The most common types of optical sensors are detecting color or fluorescent changes resulting from intermolecular interactions with the chromophore/fluorophore and the analyte. Unique optical "fingerprint" for a target analyte can be produced using new digital imaging techniques coupled with array-based methods employing chemically different groups of cross-reactive sensing elements [14].

2.2 Optical sensors

Electromagnetic radiation interacts with matter over a wide range of frequencies and in a highly specific manner. These interactions are studied and used in spectroscopy to provide information such as electronic structure of atoms or dynamics of polymeric chains. Generally, electromagnetic radiation passes through the sample and change in some optical parameter is used to evaluate the interaction of the electromagnetic radiation with the sample and associated with the analyte concentration [15].

Optical chemical sensors may be classified in general as direct sensors and reagent-mediated sensors. Analyte is directly detected via basic optical property such as absorption or luminescence in direct optical sensors. However, in reagent-mediated sensors analyte concentration is detected using the difference originating from the optical response of an intermediate, generally a dye molecule specific to an analyte. Reagent-mediated technique is especially useful for analytes which lack appropriate intrinsic optical feature [16].

Infrared, visible, or ultraviolet light is used via optical chemical sensors to analyze chemical interactions. Optical sensors are classified according to the structure of transduction mechanism as absorption, diffraction, refraction, and scattering, etc. Light source, a device for wavelength selection such as filters or monochromator, sample cell and detector are the main components of colorimetric and fluorometric sensors [14].

Fluorescence has better sensitivity and thus, it is the most frequently used spectroscopic method. When an excited electron from a molecule relaxes to the ground state emits a photon from the excited singlet state and this event is termed as fluorescence [17]. Several parameters such as intensity, emission, excitation, anisotropy, quantum yield, fluorescence decline, and anisotropy are favored for data analysis in fluorescence measurements.

Quantitative measurement of UV-visible absorbance/reflectance in other words colorimetry is one of the earliest analytical techniques. This sensing is a completely direct technique. Various portable and efficient detection methods emerged after the advent of universal digital color imaging. Although, traditional three-channel visible range (red, green and blue, RGB) is used by many colorimetric sensors, nonvisible wavelengths or full wavelength range utilizing hundreds of color channels can also be used in sensors. Few distinct wavelengths or highest peaks in UV-visible spectra are selected and analyzed in colorimetric methods to ease instrumentation and data analysis.

Partially selective interaction with analytes and providing reports via chromophore or fluorophores due to such interactions are the essential requirements of the optical sensor design. Optical sensors may be semi-selective or highly selective. These sensors must involve a probe unit that supply molecular recognition with the analyte to a certain extent, and a reporter unit for optical transduction. Both intentions can be provided by a molecule [14].

2.3 Arrays

Colorimetric arrays offer speed, easy operation due to the usage of solid strips, and portable and affordable detection of unknown analytes and thus, are advancing tools for portable detection and identification. They are attractive since most of the available techniques are time-consuming, hypothetical, and necessitate technical know-how and usage of particular chemicals and instruments [18].

Synthetic receptor in differential sensing technique does not have to be specific or very selective for the target analyte binding. This technique enables unselective and differential interactions with the target analytes, and identifies each of the analytes to different extents by using cross-reactive receptors. Thus, collection of fingerprint responses is obtained. These fingerprint responses lead to a different pattern for a single analyte or multi-analyte containing mixture [13].

3. Fabrication methods

Generally, solid arrays are fabricated by incorporating sensing dyes and pigments onto different solid supports. This type of arrays does not require liquid handling and pipetting of sensors and analytes, compared with liquid ones [18]. Polymer, nylon, paper, and nitrocellulose are widely used as supports and several immobilization methods were applied for improved selectivity [19]. Although pigment or immobilized soluble dyes have specially developed for bulk films, monoliths, sol–gel matrices, and plasticized polymer films, solid-state pigment is preferred to a soluble dye for obtaining more robust colorimetric array [16, 20].

Sol-gel method is used for formation of a wide range of material as monoliths, thin films, fibers, and powders with controlled physical and chemical properties such as mechanical strength, transparency, pore size and distribution. Nowadays, the sol-gel technology has also presented a new way of development of biosensors. The sol-gel derived materials provide an encapsulation or immobilization matrices for indicators, sensor molecules, or enzymes [21]. Sol-gel method is used to functionalizing dye molecules on the top of carbon nanotube (CNT) networks to detect various hazardous gases. The fabrication process of a CNT-dye hybrid gas sensor consist of three basic steps as coating of polietilentereftalat (PET) film by a CNT network via a drop-cast method, fabrication of electrodes by using copper tapes and coating of dye-functionalized sol-gel matrix onto the CNT network [22]. Digital image technology can be easily adapted for detection of color intensity [23]. Choodum et al. [24] reported sol-gel process for developing colorimetric sensors for methamphetamine (MA) analysis. In sol-gel procedure, MA sensor was prepared by entrapping Simon's reagent into the sol–gel solution and transferring to the flat cap micro-polymerase chain reaction (PCR) tube. After waiting for 30 min, sol-gel MA sensor was ready to use and the sensor was adapted as an application on a mobile phone [24]. Sol-gel method can be modified in order to obtain the nanoporous structure. Bang et al. [20] reported an ultrasonic based spray aerosol–gel technique for synthesis spheres in micro scale of nanoporous pigment inks as colorimetric sensor arrays. They designed the colorimetric sensor arrays by printing inks of these chemically responsive pigments. Toxic aliphatic amines selected as the analyte for testing the effectiveness of the colorimetric sensor arrays. Experimental procedure was conducted in four consecutive steps as: (1) Synthesis of chemically responsive pigments by ultrasonic spray aerosol–gel method, (2) Dispersion of the nanoporous silica-dye pigments in deionized water and 2-methoxyethanol, (3) Sensing test for detection of aliphatic amines using a static cell [20]. In addition to this, molecular imprinting technology was combined with sol-gel method for developing nicotinic acid fluorescent opto-sensing system. It was prepared fluorescent probe as carbon-dots in the silica nanoparticles and nicotinic acid in spiked human urine samples were determined in the 0.5 to 10.5 µM concentration range, with a 12.6 nM detection limit [25].

3.1 Thin-films

Generally, a thin film is defined as a layer of material at different thicknesses from nanometer to micrometers. Thin films are more suitable for especially gas-sensors because the sensing properties are related to the surface, where the gases are adsorbed and the surface reactions occur [26]. Sol-gel derived xerogel method is used for fabrication of fluorometric sensors. Chodavarapu et al. [27] investigated the sol-gel derived xerogel thin-film elements for fluorometric oxygen (O_2) sensor system which was important for various biochemical, food-beverage, and environmental applications. Firstly, the sol stock solution was prepared and doped with ruthenium containing ethanol solution. To obtain sensor films, doped sol mixture was pipetted onto a clean glass microscope slide and cured in the dark at room temperature for ten days. It was concluded that, sol-gel derived xerogel O_2 films can be further modified for developing smart sensors with high accuracy [27].

Colored thin films include cellulose nanocrystal (CNC) designed for colorimetric detection of aldehyde gases. Hydrolysis, centrifugation, and dialysis were performed in sequence for obtaining the CNC from pulp-based filter paper. Pulling process was performed to obtain the CNC film after the deposition of CNCs on the silicon substrate. Then, CNC film functionalized with amine groups and became ready for aldehyde detection by static and dynamic experiments to monitor and analyze the color changes in the films. The films have been shown to distinguish between different aldehyde gases and other interacting molecules [28]. Spin coating method used for preparation of thin films for colorimetric gas detection. Thin films at different thickness were fabricated from ultra-small p-toluenesulfonic acid (PTSA)-coated TiO_2 nanoparticles on a Si substrate. It was demonstrated that the proposed colorimetric sensitive sensor was very simple and cheap to manufacture [29]. In addition to this, it was shown that polymerization of electroactive precursors directly on the indium tin oxide electrode developed a new fluorescent film sensor. The detection ability of films for different metal ions were investigated by using the photoluminescence spectra and the results underline that film sensors are suitable for the selective and sensitive detection of Fe^{3+} ions [30]. It was reported that chemical and physical interaction between the analyte and polymer affect the response characteristics of a thin-film polymer fluorescence sensor [31]. Several types of thin films were reported as colorimetric and fluorometric sensor as curcumin nanoparticle doped starch thin film for detection of boron [32]; polyfluorene and bromocresol green assembled in thin films for a fluorescence gas sensor [33].

3.2   Polymers

In solid-state sensor fabrication, immobilization is a required step to improve the reliability by preventing the leakage of reagents. Sheet or film types of active sensing polymers can be deposited on several types of substrates for inorganic solid-state devices. The sensing effect in polymer can be stated in different materials such as dielectrics, conductive or other composites, electrolytes, etc. [34]. Zhang et al. [35] applied filtration-based enrichment method for preparing the postage stamp-sized array sensor for the sensitive screening test of heavy-metal ions. Hg^{2+}, Pb^{2+}, Ag^+, Ni^{2+}, Cu^{2+}, Zn^{2+}, and Co^{2+} heavy-metal ions at the wastewater was detected by polymer type sensor including heterocyclic dyes. After dissolving of heterocyclic dyes in the polymer formulation, polymer solutions were printed onto the hydrophilic spots. Then obtained postage stamp-sized filter paper then stored in a nitrogen-filled desiccator at 30% relative humidity for at least 3 days. Fabricated sensor showed sensitive detection of seven heavy metal ions even down to 0.05 mg l^{-1} (Hg^{2+}) [35].

Electrospinning techniques are preferred to be used for the fabrication of fluorescent and colorimetric electrospun nanofibers. Electrospun nanofibers (ENF) with large surface area, high aspect ratio, and porous structure have been introduced to fabrication and resulting electrospun sensors were with high sensitivity, good recovery, and great sensory response [36]. The major advantage of using ENF in sensors is the elimination of contamination of the detection solution and post-treatment after the detection process can be easily performed [37, 38]. Ma et al. [39] reported the fabrication of fluorescent nanofibrous membrane for mercuric ion (II) detection. Polyacrylonitrile (PAN) nanofibrous membranes were prepared by electrospinning method at parameters as 13 to 16 kV high voltage, 1.0 mL/h feed rate, 10 cm distance, 22–25 °C temperature and 30–35% relative air humidity. Fluorescent nanofibrous membrane (FNFM) was prepared via immobilization of dithioacetal-functionalized perylenediimide (DTPDI) on the surface of electrospun PAN nanofibrous membrane. Scanning electron microcopy (SEM) images clearly shown the strong red fluorescence emission of the nanofibers could be observed even by naked eyes, which only comes from DTPDI. As results of sensing experiments, FNFM reveals excellent sensitivity and selectivity for Hg^{2+}. The detection limit of FNFM can reach as low as 1 ppb [39].

Several natural and synthetic polymers used for fabrication of nanofibers for colorimetric and fluorometric sensor applications. 2-(5-Bromo-2-pyridylazo)-5-(diethylamino) phenol (Br-PADAP) embedded in cellulose acetate electrospun nanofibers for visual uranyl recognition [40], conjugated polymer-embedded electrospun fibers for volatile organic compounds detection [41], nanogold probes immobilized polyamide-6/nitrocellulose nano-fibers/nets for lead (II) assay utilizing [42]; polydiacetylenebiothiol sensor based on decomposition of a pyridine-mercury complex [43], porphyrin-containing polyimide nanofibrous membrane for pyridine vapor detection [44].

3.3 Monoliths

Monoliths are highly porous inorganic or organic materials can be designed for immobilization of sensing molecules for sensors. Generally, monolith supports are prepared based on polymerization techniques [45]. Chuag et al. [46] reported surface-modified porous monoliths and gold nanoparticles (AuNPs) for colorimetric immune sensing. Porous monoliths were prepared as solid support by photopolymerization *in situ* within a silica capillary. AuNPs are designed for colorimetric immunoassays without the need for enzyme, substrate and sophisticated equipment. Experimental procedure consist of several steps as preparation of AuNPs, formation of photopolymerized monoliths, modification of monoliths with antibodies and formation of antibody conjugated AuNP probes [46]. Silica monoliths can be organized with nanostructured mesopores with promising physical properties to meet the sensitive detection of analytes. Micro-emulsion technique, using trimethyl benzene as the solvent, used for direct templating synthesis of a mesoporous worm-like silica monolithic material using F127-a triblock copolymer. 4-dodecyl-6-(2-pyridylazo)-phenol (DPAP) was immobilized into the monolith pores for obtaining the solid-state naked-eye colorimetric ion-sensors. Pb^{2+} ions at ppb level were tested with DPAP anchored silica monolithic sensor. It was concluded that pH is the key parameter affect the exibility of molecular receptor [47]. Nanoporous monolithic SiO_2 doped with fluoral-P, were prepared by the sol-gel method for colorimetric detection of formaldehyde. It was aimed to fabricate the sensor system as a pollution-warning kit for homes. Detection system worked based on pumping of air by micro pump to flow cell which contain the monolith and measure absorbance signal with the miniature spectrophotometer. After the processing of signal with a computer, the concentration of formaldehyde is measured [48]. Moreover, monolith supports are selected for fabrication of colorimetric and fluorometric sensor for selective detection of monochloramine in air and in water [49], enhanced Co(II) ion monitoring in solution [50], Ca^{2+} based on an induced change in the conformation of sol–gel entrapped parvalbumin [51].

3.4 Pellets

Pellets as solid support for sensors are preferred for easier storage and use for practical applications. Moreover, pellets in sensor system shown long-term stability, reproducibility, and regeneration option and they can be useful for the fabrication of inexpensive sensor devices even for naked-eye detection [52]. Porous γ-alumina pellets were used for the design of the solid state sensors for direct colorimetric detection of CO_2 in the gas phase. Pellets were chemically impregnated with the amino alcohols (triethanolamine) as CO_2 and SO_2 responsive. Also, pH-indicator dye, e.g. cresol red were immobilized on porous γ-alumina pellets. It was concluded that sensors show low detection level of CO_2 and SO_2 present in air [53]. Shenashen et al. [54] used the aluminosilica pellets for the fabrication of micro-object meso-sensor for visual monitoring and also removal of copper, cadmium, and mercury ions from water. Aluminosilica pellets functionalized by porphyrinic chelating ligand were designed to be pH-dependent micro-object optical sensor for detection of metal ions in drinking water and red blood cells [54].

4. Applications

4.1 Toxic chemicals and heavy metals in water

Owing to the important roles of toxic chemicals in environments, recognition and sensing of these have been taken into attention. One of the most dangerous chemicals for living organism is cyanide that is allowed to 1.9 μM in drinking water due to its high toxicity as its absorption in lungs, gastrointestinal tract and skin can lead to vomiting, convulsions, loss of consciousness, and ultimately death. Generally, metallurgical process led to contaminate of these chemical into environment. Optical chemo-sensors for these types of chemicals (CN⁻) work via hydrogen-bonding interactions of metal complex displacement mechanism. Colorimetric naked-eye detection method is being important day by day due to make the detection easily and fast without use of expensive detections. Besides, several detectors have been improved for toxic substance, application of cyanide is limited [55-58].

Detection of cyanide ions can be performed by two types of sensors: (i) conjugated pyrene-benzothiazole system and (ii) nitrophenol-benzothiazole system. The first one only detected cyanide ions in 50% aqueous systems. Both of them cannot detect cyanide effectively in presence of more than 50% water in the medium [59, 60]. Dong et al. [56] improved reversible colorimetric chemo-sensor derived from quinoline-2-carbaldehyde and 2,4-dinitrophenylhydrazine for the detection of cyanide in dimethylsulfoxide(DMSO)/water medium based on color change from yellow to wine red. The significant change in bathochromic shift enabled the sensing of cyanide ions by naked eye. It was also reported as reversible detectors containing Ag^+ or Cu^{2+} reagents [56]. Kim et al. [61] improved conjugated naphthoquinone-benzothiazole based detector for the detection of cyanide in presence of dimethylformamide (DMF)-water colorimetrically and fluorimerically [61].

Heavy metals such as Ag, Cu, Co, Hg, etc. are one of the most toxic material group for living organisms. Accumulation of these in human body can be resulted with inactivation of enzymes and lead to cancer and neuroanatomical changes. The World Health Organization (WHO) has been fixed the limits for the heavy metals in drinking water. The colorimetric and fluorometric detection of heavy metals have been gain attention due to direct and practical detection by naked eyes [62-64].

Gao et al. [63] prepared a highly selective and sensitive reusable (more than 50 times) thiadiazole-functionalized polyacrylonitrile fiber qualitative colorimetric sensor for the detection of Ag^+ ions in presence of other heavy metal ions. The detection of Ag^+ ions could be performed by color change from the yellow to red-brown in the wide range of pH (3-12.9) with 11 ppb detection limit. They reported that fiber type detectors enabled excellent selectivity, lower detection limit, outstanding absorption capacity and reusability to detect and remove Ag^+ from water [63].

Ghodake et al. [65] reported spherical gallic acid-functionalized Ag nanoparticles (NPs) for the colorimetric and spectrophotometric detection of Al^{3+}in an aqueous medium. The formation of chelating complex between hydroxyl (OH^-) and Al^{3+} ions led to color change from yellow to red. The detection limit of the nanoparticles were determined as 0.55 μM (spectral) and 0.92 μM (visual) which are lower than the drinking water limits (7.41 μM) [65]. Liu et al. [66] developed label-free colorimetric probe based on catalytic etching of AuNPs for the colorimetric sensing of Cu^{2+}. A visible color change from red to colorless with a sensitivity limit of detection (LOD) 5.0 nM or 0.32 ppb and a selective (at least 100 fold over other metal ions except for Pb^{2+} and Mn^{2+}) way for the detection of Cu^{2+} (linear range,10–80 nM) is obtained [66]. A flow batch type analysis has been used for the colorimetric detection of Cu^{2+}via AgNPs. Peng et al.[67] tested the NPs in presence of several metallic ions such as Fe^{3+}, Mn^{2+}, Co^{2+}, Ni^{2+}, Zn^{2+}, Pb^{2+}, Ba^{2+}, Cd^{2+}, Bi^{3+}, Sb^{2+}, As^{3+}, Hg^{2+}, Cr^{3+} and K^+. The improved method was reported to be appropriate for several water samples taken from lake, tap, rain and bottled, food packaging with

0.9954 coefficient in 0.5-35 $\mu g.l^{-1}$ range [67]. Cao et al. [68] provided a portable colorimetric method to detect Cu^{2+} ion in drinking water by using red beet pigment and smartphone. The detection limit of the system reported was in 4-20 μM range. The red beet pigment was interacted with Cu^{2+} ions and the color changed to blue. The Android application was developed and successfully detected the Cu^{2+} ion in drinking water with 0.84 μM detection limit [68]. Despite the trace elements have key role such as Fe^{3+} in human body, the deficiency and excessive amount of these are not desired for healthy organism. Ultra-sensitive colorimetric sensing in human blood is an important topic. Liu et al. [69] reported a method by using Au nanorods and using H_2O_2 as an oxidation agent. The method worked in the wide range as 0.20-30 μM Fe^{3+} in blood successfully [69].

4.2 Pharmaceutical and medical

The detection of metabolites such as urea, amino acids, proteins, vitamins etc. by colorimetric and fluorescent sensors have been one of the primary interests of science. Arginine is one of the important ones, since it is a precursor of urea, ornithine etc. that influence metabolism and excess or absence of it induces life-threat. For these reasons, urgent and practical detection of this molecule is important. Naked eye-sensing of these metabolite have been reported by Schifft base ligands. The chemo-sensors based on cations and neutral molecules use ligands with metals such as Pb^{2+} etc., nanoparticles [70-72]. Ghorai et al. [73] synthesized Pb^{2+}-complex as a colorimetric-fluorescent–colorimetric sensor. The sensor showed on-off fluorescence change within 1 min by low detection limits (0.67 μM) and rapid response [73]. Peng et al. [74] investigated Ag nanoclusters (NCs) for the rapid, sensitive and selective colorimetric detection of ascorbic acid at room temperature with detection limit as low as 79.2 nM. Also these NCs gave high performance in presence of other metabolites such as glucose, dopamine, uric acid and cysteine [74]. Rostami et al. [75] prepared seed-mediated grown AgNPs for detection of ascorbic acid with 0.054 μM. Rahim et al. [76] researched the quantitative determination of pesticide Cartap (0.036–0.36 μgl^{-1}) in aqueous media and blood plasma by using polystyrene-block-poly(2-vinylpyridine)-conjugated AgNPs with LOD of 0.06 μgl^{-1} [76]. Star-shaped poly(ethylene oxide)-block-poly(caprolactone) conjugated AgNPs were also performed for cephalexin (LOD: 1.8 μM) detection in environmental, biological and pharmaceutical samples. AgNPs showed brilliant properties for the detection of pharmaceutics in different media. Their on-spot alarm property enable practical and economical determination compared the other methods [77].

4.3   Food industry

In the food industry, there are several substances and residues coming from production units or the main source of food which are very harmful for human beings. Determination of these residues generally based on the analytical techniques. Getting results by a fast and adequate way is the most curial point. The researchers have been working on the detection of the substances such as antibiotics, bacteria, pesticides, glucose, ascorbic acid, organosulfur compounds, metal and polyatomic cyanide, allergens etc. [78]. National and international organizations restricted the limits of several substances in food. As a result of these, maximum residue limits (MRLs) were announced. The detection of antibiotics is one of the prominent ones in food sector. This is a chemical substance that is used to decrease the target microorganism. Especially in milk industry, antibiotics used for the healing of mastitis of dairy cows and the residues of antibiotics are found in the milk. These residues have become important even after 2001, when the harms of antibiotics were understand for publicity and the antibiotics in the milk decreased the efficiency of fermentation process. As an alternative to conventional methods such as chromatography, microbial growth inhibition assays, colorimetric detection provides advantages such as being fast, economic etc. Zhang et al. [79] improved a highly sensitive and selective colorimetric sensing technique for antibiotics in milk by using AuNPs. The technique was reported to be sensitive for kanamycin mono sulfate, neomycin sulfate, streptomycin sulfate and bleomycin sulfate antibiotics in presence of several substrate as phenylalanine, alanine, glycerol, glucose, Mg^{2+}, Ca^{2+}, Na^+, K^+, CO_3^{2-}, SO_4^{2-}, NO_3^- and Cl^- ions [79].

The other most curial issue in food sector is freshness of the products. Especially, chicken and fish have a disadvantage with their short shelf lives. The quality of the products and freshness are the requirements for the public. Generally physical, chemical, microbiological techniques and human sensory evaluation have been used. Yet, the fast, economic and efficient techniques need to get the accurate results. Zhai et al. [80] prepared hydrogen sulphide colorimetric sensor based on gellan gum-AgNPs bionanocomposite for the detection of meat spoilage. The intelligent packing of products enable customers safe products. Their sensor has a 0.81 µM limit of detection and very sensitive results visualized by changing color from yellow to colorless. This provided useful, economic, and an user-friendly approach for food sector [80]. Also e-nose application is popular for evaluating the food odor based on volatile compounds of stale products. The hydrophobic sensor material and plate with colorimetric sensitive dyes were formed the e-nose [81]. The interaction of dye and chemical vapor of target molecule resulted with detection and evaluation of product. Chen et al. reported a study to evaluate chicken freshness based on a fabrication of dye, AdaBoost-OLDA algorithm.

They used total volatile basic nitrogen (TVB-N) content of chicken. To evaluate the sample freshness, the specific calorific fingerprint was used based on biochemical process. 12 chemically responsive dyes on C2 reverse silica-gel flat plate was used and successfully evaluated the freshness of chicken based on AdaBoost-OLDA algorithm [82]. Huo et al. [83] designed colorimetric artificial tongue and nose to determine nine different Chinese tea by using chemoresponsive dyes such as porphyrin, dimericmetalloporphyrins, metallosalophen complexes on the hydrophobic membrane. The discrimination of different tea samples was successfully achieved according to their origin and grade level using colorimetric technique [83].

Biological thiols such as cysteine, glutathione etc. are critical biomolecules for the biological systems that can cause to slow down the growing process of children, loose muscle and fat, damage of viscus. Colorimetric sensing based on nanoparticles in liquid sensory is one of the way to detect these. Ghasemi et al. [84] developed a sensitive, cheap and rapid system for detection of molecules such as cysteine, glutathione and glutathione disulphide with 10-800 μM detection range [84]. Huang et al. [85] used natural dye obtained from Arnebia Euchroma root to detect the freshness of fish under refrigeration and room conditions. Total volatile basic nitrogen (TVB-N) and viable count (TVC) were periodically monitored and the color change was observed. The indicator films provide visual and practical determination of the freshness of fish and smart food packing [85].

Cyanide (CN^-) is one of other important molecule that effect with its high toxicity and it is found in herbicides, dyes, and leather industry. Several vegetables, plants, salts and stabilizer have low concentration of CN^-. So this and its high toxicity affect several systems in human body thus its practical detection gains importance. Besides the spectroscopic methods, detection of it by fluorescent sensors based on different chemical reaction mechanisms including hydrogen bonding motifs, nucleophilic addition, supramolecular self-assembly, electron deficient alkenes have been investigated [86-91]. Niu et al. [86] developed "naked-eye" colorimetric and fluorescent "turn-on" detection technique for CN^- in food samples and living HeLa cells. Paper based colorimetric strips of 3- total base number (3TBN) sensor showed visual color and spectral changes during detection with low detection limit as 0.46 μM. The detection of CN^- carried out by nucleophilic addition reaction with CN^- on the cyanovinyl electrophilic C-C group [86].

Melamine is the one of the most common one that could be found frequently in protein rich foods. Melamine interacts with cyanuric acid and forms insoluble crystals that could damage human's viscus. Despite several analytical and instrumental methods have been improved, practical developments are welcome to fast and economic solutions. Rajat et al. studied on a strategy to detect this compound in milk samples. They preferred to use AgNPs which were functionalized by succinic acid for this aim. Succinic acid provided a formation of link with amine moieties of melamine and changed absorption enabling colorimetric determination of melamine (0.1-1.2 μM) [92].

Concluding remarks

The chapter focused fundamentals, fabrication methods, usage areas of colorimetric/ fluorometric sensor arrays and the following remarks concluded:

Feature: Main features of sensor arrays are high level discriminatory and specific responses to wide range of substrates. Progress in this chemical sensing, will present data for recognition of artificial intelligence and developing areas of machine learning.

Fabrications: The important parameters to be take into consideration of design of novel sensing devices are to achieve high stability and high sensitivity with low power and low cost. The fabricated of colorimetric/fluorometric arrays based chemical detecting platforms has been focused on immobilization or encapsulation of indicator, sensor molecules, or enzyme matrices on solid supports (thin films, polymers, fibers, monoliths, pellets) synthesized with advanced techniques. Chemically diverse chemo responsive dyes or bio-receptors can be used in conjunction with chemometric analysis to design sequences to distinguish between various substrates and their mixtures with similar structure. It is also important to integrate the sensor arrays with digital imaging technology such as digital cameras, card scanners and cell phones.

Applications: The advances in sensitive and reliable array sensing for single analyte and their mixtures have been turn them into suitable tools for gas and liquid specimens. Beside this, the main limitation of array detection is that they are not adequate for component-by-component measurement of mixtures. Array sensing provides to be a superior method for the quality control and quality assurance goals.

References

[1]   RO'Kennedy, W.J.J. Finlay, P.Leonard, S. Hearty, J. Brennan, S. Stapleton, S. Townsend, A. Darmaninsheehan, A. Baxter, C. Jones, Applications   of sensors in food and environmental analysis, in: M.K. Ram, V.R. Bhethanabotla (Eds.), Sensors for chemical and biological applications, CRC Press-Taylor and Francis Group, BocaRaton, 2010, pp. 195-232.

[2]   B.R. Eggins, Introduction chemical sensors and biosensors, in:D.J. Ando (Eds.), Analytical techniques in the sciences, John Wiley & Sons Ltd, West Sussex (2004) pp. 1-9

[3]   J.I.A. Ming-Yan, F.E.N.G.Liang, Recent progresses in optical colorimetric/fluorometric sensor array, Chinese J. Anal. Chem. 4(2013) 795-802. https://doi.org/10.1016/S1872-2040(13)60658-1

[4]   D. Prabhakaran, C.Subashini, M.A. Maheswari, Synthesis of mesoporous silica monoliths-a novel approach towards fabrication of solid-state optical sensors for environmental applications, Int. J. Nanosci. 15 (2016) 1660014. https://doi.org/10.1142/S0219581X16600140

[5]   X. Zhang, J. Yin, J. Yoon, Recent advances in development of chiral fluorescent and colorimetric sensors,Chem. Rev.114 (2014) 4918-4959. https://doi.org/10.1021/cr400568b

[6]   A. Senthamizhan, A. Celebioglu, B. Balusamy, T. Uyar, Immobilization of gold nanoclusters inside porous electrospun fibers for selective detection of Cu(II): a strategic approach to shielding pristine performance, Sci. Rep. 5 (2015) 15608-15619. https://doi.org/10.1038/srep15608

[7]   S.H. Lim, J.W. Kemling, L. Feng, K.S. Suslick, A colorimetricsensor array of porous pigments, Analyst. 134 (2009) 2453-2457. https://doi.org/10.1039/B916571A

[8]   J. Zhang, F. Cheng, J. Li, J.J. Zhu, Y. Lu, Fluorescent nanoprobes for sensing and imaging of metal ions: recent advances and future perspectives, Nano. Today. 11 (2016) 309–329. https://doi.org/10.1016/j.nantod.2016.05.010

[9]   A. Escudero, A.I. Becerro, C. Carrillo-Carrión, N.O. Núñez, M.V. Zyuzin, M. Laguna, D. González-Mancebo, M. Ocaña, W.J. Parak, Rare earth based nanostructured materials: synthesis, functionalization, properties and bioimaging and biosensing applications, Nanophotonics. 6(2017) 881-921. https://doi.org/10.1515/nanoph-2017-0007

[10]   N. Aliheidari, N.Aliahmad, M. Agarwal, H. Dalir, Electrospun nanofibers for label-free sensor applications, Sensor. 19 (2019) 3587-3614. https://doi.org/10.3390/s19163587

[11]   K.J. Albert, N.S. Lewis, C.L. Schauer, G.A. Sotzing, S.E. Stitzel, T.P. Vaid, D.R. Walt, Cross-reactive chemical sensor arrays, Chem. Rev. ,100 (2000) 2595-2626. https://doi.org/10.1021/cr980102w

[12]   A.E. Fenster, D.N. Harpp, J.A. Schwarcz, A useful model for the" lock and key" analogy, J. Chem. Edu. 61 (1984) 967. https://doi.org/10.1021/ed061p967

[13]   S.F. Wong, S.M. Khor, State of the art of differential sensing techniques in analytical sciences, TrAC, 114(2019) 108-115.https://doi.org/10.1016/j.trac.2019.03.006

[14]   Z. Li, J.R.Askim, K.S. Suslick, The optoelectronic nose: colorimetric and fluorometric sensorarrays,Chem. Rew. 119(2018) 231-292. https://doi.org/10.1021/acs.chemrev.8b00226

[15]   J. Janata, Optical sensors, in:. Janata (Eds.), Principles of chemical sensors, Springer Science Business Media, New York, 2009, pp.267-311.

[16]C. McDonagh, C.S. Burke, B.D. MacCraith, Optical chemicalsensors, Chem. Rew. 108 (2008) 400-422. https://doi.org/10.1021/cr068102g

[17]   J.R. Lakowicz, Introductionto fluorescence, in: Joseph R. Lakowicz (Eds.), Principles of fluorescence spectroscopy, Springer Science Business Media, New York,2006, pp. 1-26.

[18]   M.J. Kangas, R.M. Burks, J. Atwater, R.M. Lukowicz, P. Williams, A.E. Holmes, Colorimetric sensor arrays for the detection and identification of chemical weapons and explosives, Crit. Rev. Anal. Chem. 47 (2017) 138-153. https://doi.org/10.1080/10408347.2016.1233805

[19]   T. Soga, Y. Jimbo, K. Suzuki, D. Citterio, Inkjet-printed paper-based colorimetric sensor array for the discrimination of volatile primary amines, Anal.Chem. 85(2013) 8973-8978. https://doi.org/10.1021/ac402070z

[20]   J.H. Bang, S.H. Lim, E. Park, K.S. Suslick, Chemically responsive nanoporous pigments: colorimetric sensor arrays and the identification of aliphatic amines, Langmuir 24 (2008) 13168-13172. https://doi.org/10.1021/la802029m

[21]   H. Podbielska, A. Ulatowska-Jarża, G. Müller, H.J. Eichler, Sol-gels for optical sensors. in: Baldini F., Chester A., Homola J., Martellucci S. (Eds), Optical chemical sensors. Springer, Dordrecht, 2006, pp. 353-385

[22]   J. Kim, H. Yoo, V.A.P. Ba, N. Shin, S. Hong, S., Dye-functionalized sol-gel matrix on carbon nanotubes for refreshable and flexible gas sensors, Sci. Rep. 8(2018) 1-8. https://doi.org/10.1038/s41598-018-30481-y

[23]   A. Choodum, K. Parabun, N. Klawach, N.N. Daeid, P. Kanatharana, W. Wongniramaikul, Real time quantitative colourimetric test for methamphetamine detection using digital and mobile phone technology, Forensic. Sci. Int. 235 (2014) 8-13. https://doi.org/10.1016/j.forsciint.2013.11.018

[24]   A. Choodum, P. Kanatharana, W. Wongniramaikul, N. NicDaeid, A sol–gel colorimetric sensor for methamphetamine detection. Sens. Actuators. B Chem. 215 (2015) 553-560. https://doi.org/10.1016/j.snb.2015.03.089

[25]   P. Zuo, J. Gao, J. Peng, J. Liu, M. Zhao, J. Zhao, P. Zuo, H. He, A sol-gel based molecular imprint incorporating carbon dots for fluorometric determination of nicotinic acid, Microchim. Acta. 183(2016) 329-336. https://doi.org/10.1007/s00604-015-1630-5

[26]   P.P. Sahay, R.K. Nath, Al-doped ZnO thin films as methanol sensors, Sens. Actuators. B Chem. 134 (2008) 654-659. https://doi.org/10.1016/j.snb.2008.06.006

[27]   V.P. Chodavarapu, D.O. Shubin, R.M. Bukowski, A.H. Titus, A.N. Cartwright, F.V. Bright, CMOS-based phase fluorometric oxygen sensor system. IEEE. T. Circuits-I,54 (2007) 111-118. https://doi.org/10.1109/TCSI.2006.888680

[28]   W. Song, J.K. Lee, M.S. Gong, K. Heo, W.J. Chung, B.Y. Lee, Cellulose nanocrystal-based colored thin films for colorimetric detection of aldehyde gases, ACS Appl. Mater. Interfaces, 10 (2018) 10353-10361. https://doi.org/10.1021/acsami.7b19738

[29]   U. Joost, A. Šutka, M. Visnapuu, A. Tamm, M. Lembinen, M., Antsov, K. Utt, K., Smits, E. Nõmmiste, V. Kisand, Colorimetric gas detection by the varying thickness of a thin film of ultrasmall PTSA-coated TiO$_2$ nanoparticles on a Si substrate, Beilstein. J.Nanotechnol. 8 (2017) 229-236. https://doi.org/10.3762/bjnano.8.25

[30]   P. Li, C. Ji, H. Ma, H., M. Zhang, Y. Cheng, Development of fluorescent film sensors based on electropolymerization for iron(III) ion detection, Chem.Eur. J. 20(2014) 5741-5745. https://doi.org/10.1002/chem.201304364

[31]   Y. Liu, R.C. Mills, J.M. Boncella, K.S. Schanze,. Fluorescent polyacetylene thin film sensor for nitro aromatics, Langmuir17 (2001) 7452-7455. https://doi.org/10.1021/la010696p

[32]   C. Boonkanon, K. Phatthanawiwat, W.Wongniramaikul, A. Choodum, Curcumin nanoparticle doped starch thin film as a green colorimetric sensor for detection of boron, Spectrochim. Acta. A. 224 (2020) 117351. https://doi.org/10.1016/j.saa.2019.117351

[33]   M.G. Guillén, F.Gámez, T. Lopes-Costa, J. Cabanillas-González, J.M. Pedrosa, A fluorescence gas sensorbased on Förster resonance energy transfer between polyfluorene and bromocresol green assembled in thin films, Sens. Actuators. B. Chem. 236 (2016) 136-143.https://doi.org/10.1016/j.snb.2016.06.011

[34]   G. Harsányi, Sensing effects and sensitive polymers, in: G. Harsányi, (Ed.) Polymer films in sensor applications, Technomic Publishing Company Inc., Pennsylvania,1995, pp. 93-261.

[35]   Y. Zhang, X. Li, H. Li, M. Song, L. Feng, Y. Guan, Postage stamp-sized array sensor for the sensitive screening test of heavy-metal ions, Analyst. 139 (2014)4887-4893. https://doi.org/10.1039/c4an01022a

[36]N. Aliheidari, N. Aliahmad, M. Agarwal, H. Dalir, Electrospun nanofibers for label-free sensor applications, Sensors. 19 (2019) 3587-3614. https://doi.org/10.3390/s19163587

[37]S.J. Choi, L. Persano, A. Camposeo, J.S. Jang, W.T. Koo, S.J. Kim, H.J.Cho, I.D. Kim, D. Pisignano, Electrospun nanostructuresfor high performance chemiresistiveand optical sensors, Macromol. Mater. Eng. 302 (2017) 1-37. https://doi.org/10.1002/mame.201600569

[38]   N. Zhang, R. Qiao, J. Su, J. Yan, Z. Xie, Y. Qiao, X. Wang, J. Zhong, Recent advances of electrospun nanofibrous membranes in the development of chemosensors for heavy metal detection, Small. 13 (2017) 1604293-1604311. https://doi.org/10.1002/smll.201604293

[39]   L. Ma, K. Liu, M. Yin, J. Chang, Y. Geng,K. Pan, Fluorescent nanofibrous membrane (FNFM) for the detection of mercuric ion (II) with high sensitivity and selectivity, Sensor. Actuat. B-Chem. 238 (2017) 120-127. https://doi.org/10.1016/j.snb.2016.07.049

[40]   L. Hu, X.W. Yan, Q. Li, X.J. Zhang, D. Shan, Br-PADAP embedded in celluloseacetate electrospun nanofibers: Colorimetric sensor strips for visual uranyl recognition, J. Hazard. Mater. 329 (2017) 205-210. https://doi.org/10.1016/j.jhazmat.2017.01.038

[41]   J. Yoon, S.K. Chae, J.M. Kim, Colorimetric sensors for volatile organic compounds (VOCs) based on conjugatedpolymer-embedded electrospun fibers, JACS. 129(2007) 3038-3039. https://doi.org/10.1021/ja067856

[42] Y. Li, Y. Si, X. Wang, B. Ding, G. Sun, G., Zheng, W. Luo, J. Yu, Colorimetric sensor strips for lead(II) assay utilizing nano gold probes immobilized polyamide-6/nitrocellulose nano-fibers/nets, Biosens. Bioelectron. 48 (2013) 244-250. https://doi.org/10.1016/j.bios.2013.03.085

[43]   J.P. Lee, F. Jannah, K. Bae,J.M. Kim, A colorimetric and fluorometric polydiacetylene biothiol sensor based on decomposition of a pyridine-mercury

complex, Sensor. Actuat. B-Chem. (2020) 127771.
https://doi.org/10.1016/j.snb.2020.127771

[44]   Y. Lv, Y. Zhang, Y. Du, J. Xu, J. Wang, A novel porphyrin-containing polyimide
nanofibrous membrane for colorimetric and fluorometric detection of pyridinevapor,
Sensors. 13 (2013) 15758-15769. https://doi.org/10.3390/s131115758

[45]   R.M. Kröll, N. Schuler, S. Lubbad, M.R. Buchmeiser, A ROMP-derived, polymer-
supported chiral Schrock catalyst for enantioselective ring-closing olefin metathesis,
Chem.Com. 21 (2003) 2742-2743. https://doi.org/10.1016/j.polymer.2007.02.045

[46]   S.H. Chuag, G.H. Chen, H.H. Chou, S.W. Shen, C.F. Chen, Accelerated
colorimetric immunosensing using surface-modified porous monoliths and gold
nanoparticles, Sci. Technol. Adv. Mater.14(2013) 044403-
044409.https://doi.org/10.1088/1468-6996/14/4/044403

[47]   D. Prabhakaran, C. Subashini, M.A. Maheswari, Synthesis of mesoporous silica
monoliths-a novel approach towards fabrication of solid-state optical sensors for
environmental applications, J. Nanosci. 15 (2016) 1660014.
https://doi.org/10.1142/S0219581x16600140

[48]   S. Mariano, W. Wang, G. Brunelle, Y. Bigay, T.H. Tran-Thi, Colorimetric
detection of formaldehyde: A sensor for air quality measurements and a pollution-
warning kit for homes,Sensordevices. (2010) 80-83.
https://doi.org/10.1109/sensordevices.2010.18

[49]   T.H. Nguyen, L. Mugherli, C. Rivron, T.H. Tran-Thi, Innovative colorimetric
sensors for the selective detection of monochloramine in air and in water, Sensor.
Actuat. B-Chem. 208 (2015) 622-627. https://doi.org/10.1016/j.snb.2014.10.108

[50]   S.A.El-Safty, Functionalized hexagonal mesoporous silica monoliths with
hydrophobicazo-chromophore for enhanced Co(II) ion monitoring. Adsorption,
15(2009) 227-239. https://doi.org/10.1007/s10450-009-9171-z

[51]   K. Flora,J.D. Brennan, Fluorometric detection of Ca^{2+} based on an induced change
in the conformation of sol−gel entrapped parvalbumin. Anal. Chem. 70 (1998) 4505-
4513. https://doi.org/10.1021/ac980440x

[52]   S.A. El-Safty, M.A. Shenashen, A. Shahat, Tailor made micro object optical
sensor based on mesoporous pellets for visual monitoring and removal of toxic metal
ions from aqueous media, Small. 9 (2013) 2288-2296.
https://doi.org/10.1002/smll.201202407

[53] C. Chatterjee, A. Sen, Sensitive colorimetric sensors for visual detection of carbon dioxide and sulfur dioxide, J. Mater. Chem. A 3 (2015) 5642-5647. https://doi.org/10.1039/C4TA06321J

[54] M.A. Shenashen, E.A. Elshehy, S.A. El-Safty, M. Khairy, Visual monitoringand removal of divalent copper, cadmium, and mercury ions from water by using mesoporous cubic Ia3d aluminosilica sensors, Sep. Purif.Technol. 116 (2013) 73-86. https://doi.org/10.1016/j.seppur.2013.05.011

[55] M.E. Moragues, R. Martínez-Máñez, F. Sancenón, Chromogenic and fluorogenic chemosensors and reagents for anions, Chem. Soc. Rev. 40 (2011) 2593–2643. https://doi.org/10.1039/C0CS00015A

[56] Z.M. Dong , W.Wang, Y.B. Wang, J.N. Wang, L.Y. Qin, Y. Wang, A reversible colorimetric chemosensor for "naked eye" sensing of cyanide ion in semi-aqueous solution, Inorg. Chim. Acta. 461 (2017) 8-14. https://doi.org/10.1016/j.ica.2017.02.004

[57] Y.Y. Guo, X.L. Tang, F.P. Hou, J. Wu, W. Dou, W.W. Qin, J.X. Ru, G.L. Zhang, W.S. Liu, X.J. Yao, A reversible fluorescent chemosensor for cyanide in 100% aqueous solution, Sens.Actuators. B 181 (2013) 202–208. https://doi.org/10.1016/j.snb.2013.01.053

[58] S. Goswami, A. Manna, S. Paul, K. Aich, A.K. Das, S. Chakraborty, Highly reactive (<1 min) ratiometric probe forselective ''naked-eye'' detection of cyanide in aqueousmedia, Tetrahedron Lett. 54 (2013) 1785–1789. https://doi.org/10.1016/j.tetlet.2012.12.092

[59] J.J. Li, W. Wei, X.L. Qi, G. Zuo, J. Fang, W. Dong, Highly selective colorimetric/fluorimetric dual-channel sensorfor cyanide based on ICT off in aqueous solution, Sens.Actuators. B Chem. 228 (2016) 330–334. https://doi.org/10.1016/j.snb.2016.01.055

[60] J.J. Li, X.L. Qi, W. Wei, Y. Liu, X. Xu, Q. Lin, W. Dong, A donor-two-acceptor sensor for cyanide detection in aqueous solution, Sens. Actuators. B-Chem. 220(2015) 986–991. https://doi.org/10.1016/j.snb.2015.06.042

[61] I.J. Kim, M. Ramalingam, Y.A. Son, A reaction based colorimetric chemosensor for the detection of cyanide ion in aqueous solution, Sensor. Actuat. B Chem. 246 (2017) 319-326. https://doi.org/10.1016/j.snb.2017.02.015

[62] A.W. Andren, D.E. Armstrong, The environmental chemistry and toxicology of silver, Environ. Toxicol. Chem. 18 (1999) 1–2. https://doi.org/10.1002/etc.5620180101

[63]   R. Gao, G. Xu, L. Zheng, Y. Xie, M. Tao, W. Zhang, A highly selective and sensitive reusable colorimetric sensor for Ag^+ based on thiadiazolefunctionalizedpolyacrylonitrilefiber, J. Mater.Chem. C. 4 (2016) 5996-6006. https://doi.org/10.1039/C6TC00621C

[64]   S.C. Bondy, Low levelsof aluminum can lead to behavioral and morphological changes associated with alzheimer's disease and age-related neurodegeneration, Neurotoxicology. 52 (2016) 222-229. https://doi.org/10.1016/j.neuro.2015.12.002

[65]   G. Ghodake, S. Shinde, A. Kadam, R.G. Saratale, G.D. Saratale, A. Syed, O. Shair, M. Alsaedi, D.Y. Kim, Gallic acid-functionalized silver nanoparticles as colorimetric and spectrophotometric probe for detection of Al^{3+} in aqueous medium, J. Ind. Eng. Chem. 82 (2019) 243-253. https://doi.org/10.1016/j.jiec.2019.10.019

[66]   R. Liu, Z. Chen, S.Wang, C. Qu, L. Chen, Z. Wang, Colorimetric sensing of copper(II) based on catalytic etching of gold nanoparticles, Talanta. 112 (2013) 37-42. https://doi.org/10.1016/j.talanta.2013.01.065

[67]   J. Peng, G. Liu, D. Yuan, S. Feng, S. T. Zhou,. A flow-batch manipulated Ag NPs based SPR sensor for colorimetric detection of copper ions (Cu^{2+}) in water samples, Talanta 167 (2017) 310-316. https://doi.org/10.1016/j.talanta.2017.02.015

[68]   Y. Cao, Y. Liu, F. Li, S. Guo, Y. Shui, H. Xue, L. Wang, Portable colorimetric detection of copper ion in drinking water via red beet pigment and smartphone, Microchem. J. 150 (2019) 104176-104182. https://doi.org/10.1016/j.microc.2019.104176

[69]   J.M. Liu, X.X. Wang, L. Jiao, M.L. Cui, L.P. Lin, L.H. Zhang, S.L. Jiang, Ultra-sensitive non-aggregation colorimetric sensor for detection of iron based on the signal amplification effect of Fe^{3+} catalyzing H_2O_2 oxidize gold nanorods, Talanta. 116 (2013)199-204. https://doi.org/10.1016/j.talanta.2013.05.024

[70]   A. Ghorai, J. Mondal, R. Chandra, G. K. Patra, A reversible fluorescent-colorimetric imino-pyridyl bis-Schiff base sensor for expeditious detection of Al^{3+} and HSO_3^- in aqueous media, Dalton. Trans. 44 (2015) 13261-13271. https://doi.org/10.1039/C5DT01376C

[71]   A. Ghorai, J. Mondal, R. Chandra, G. K. Patra, Exploitation of a simple Schiff baseas a ratio metric and colorimetric chemosensor for glutamicacid, Anal. Methods. 7 (2015) 8146-8151. https://doi.org/10.1039/c5ay01465d

[72]   M. Zhang, K. Gong, H. Zhang, L. Mao, Layer-by-layer assembled carbon nanotubes for selective determination of dopamine in the presence of ascorbic acid, Biosens. Bioelectron. 20 (2005) 1270-1276. https://doi.org/10.1016/j.bios.2004.04.018

[73]   A. Ghorai, J. Mondal, G.K. Patra, A new Schiff base and its metal complex as colorimetric and fluorescent–colorimetric sensors for rapid detection of arginine, New. J. Chem. 40 (2016) 7821-7830. https://doi.org/10.1039/c5nj02787j

[74]   J. Peng, J. Ling, X.Q. Zhang, L.Y. Zhang, Q.E. Cao, Z.T. Ding, A rapid, sensitive and selective colorimetric method for detection of ascorbicacid, Sensor. Actuat. B-Chem. 221 (2015) 708-716. https://doi.org/10.1016/j.snb.2015.07.002

[75]   S. Rostami, A. Mehdinia, A. Jabbari, Seed-mediated grown silver nanoparticles as a colorimetric sensor for detection of ascorbic acid, Spectrochim. Acta. A. 180 (2017) 204-210. https://doi.org/10.1016/j.saa.2017.03.020

[76]   S. Rahim, S. Khalid, M.I. Bhanger, M.R. Shah, M.I. Malik, Polystyrene-block-poly (2-vinylpyridine)-conjugated silver nanoparticles as colorimetric sensor for quantitative determination of Cartap in aqueous media and blood plasma, Sensor. Actuat. B Chem. 259 (2018) 878-887. https://doi.org/10.1016/j.snb.2017.12.138

[77]   S. Rahim, A.M. Bhayo, M.R. Shah, M.I. Malik, Star-shaped poly(ethylene oxide)-block-poly(caprolactone) conjugated silver nanoparticles: a colorimetric probe for cephalexin in environmental, biological and pharmaceutical samples, Microchem. J. 149 (2019) 104048-104057https://doi.org/10.1016/j.microc.2019.104048-104057

[78]   Y. Wang, T.V. Duncan, Nanoscale sensors for assuring the safety of food products, Curr. Opin. Biotechnol. 44 (2017) 74-86. https://doi.org/10.1016/j.copbio.2016.10.005

[79]   X. Zhang, Y. Zhang, H. Zhao, Y. He, X. Li, Z. Yuan, Highly sensitive and selective colorimetric sensing of antibiotics in milk, Anal.Chim. Acta. 778 (2013) 63-69. https://doi.org/10.1016/j.aca.2013.03.059

[80]   X. Zhai, Z. Li, J. Shi, X. Huang, Z. Sun, D. Zhang, X. Zou, Y. Sun, J. Zhang, M. Holmes, Y. Gong, A colorimetric hydrogen sulfide sensorbased on gellangum-silver nanoparticles bionanocomposite for monitoring of meat spoilage in intelligent packaging, Food. Chem. 290 (2019) 135-143. https://doi.org/10.1016/j.foodchem.2019.03.138

[81]   S. Benedetti, C. Pompei, C., S. Mannino, Comparison of an electronic nose with the sensory evaluation of foodproductsby "triangletest", Electroanal. 16 (2004) 1801-1805. https://doi.org/10.1002/elan.200303036

[82] Q. Chen, Z. Hui, J. Zhao, Q. Ouyang, Evaluation of chicken freshness using a low-cost colorimetric sensor array with AdaBoost–OLDA classification algorithm, LWT-Food Sci. Technol. 57 (2014) 502-507. https://doi.org/10.1016/j.lwt.2014.02.031

[83] D. Huo, Y. Wu, M. Yang, H. Fa, X. Luo, C. Hou, Discrimination of Chinese green tea according to varieties and grade levels using artificial nose and tongue based on colorimetric sensor arrays, Food.Chem. 145 (2014) 639-645. https://doi.org/10.1016/j.foodchem.2013.07.142

[84] F. Ghasemi, M.R. Hormozi-Nezhad, M. Mahmoudi, A colorimetricsensor array for detection and discrimination of biothiols based on aggregation of gold nanoparticles, Anal. Chim. Acta. 882 (2015) 58-67. https://doi.org/10.1016/j.aca.2015.04.011

[85] S. Huang, Y. Xiong, Y. Zou, Q. Dong, F. Ding, X. Liu, H. Li, A novel colorimetric indicator based on agar incorporated with arnebiaeuchromaroot extracts for monitoring fish freshness, Food. Hydrocolloid. 90 (2019) 198-205. https://doi.org/10.1016/j.foodhyd.2018.12.009

[86] Q. Niu, L. Lan, T. Li, Z. Guo, T. Jiang, Z. Zhao, Z. Feng, J. Xi, A highly selective turn-on fluorescent and naked-eye colorimetric sensor for cyanide detection in food samples and its application in imaging of living cells, Sensor. Actuat. B-Chem. 276 (2018)13-22. https://doi.org/10.1016/j.snb.2018.08.066

[87] G.J. Park, Y.W. Choi, D. Lee, C. Kim, A simple colorimetric chemosensor bearing a carboxylic acid group with high selectivity for CN^-, Spectrochim. Acta. A. 132 (2014) 771-775. https://doi.org/10.1016/j.saa.2014.06.001

[88] B. Shi, P. Zhang, T. Wei, H. Yao, Q. Lin, Y. Zhang, Highly selective fluorescent sensing for CN^- in water: utilization of the supramolecular self-assembly, Chem. Commun. 49 (2013) 7812-7814. https://doi.org/10.1039/C3CC44056G

[89] T. Sun, Q. Niu, Y. Li, T. Li, T. Hu, E. Wang, H. Liu, A novel oligothiophene-based colorimetric and fluorescent "turn on" sensor for highly selective and sensitive detection of cyanide in aqueous media and its practical applications in water and food samples, Sens. Actuators. B Chem. 258 (2018) 64-71. https://doi.org/10.1016/j.snb.2017.11.095

[90] Q. Niu, T. Sun, T. Li, Z. Guo, H. Pang, Highly sensitive and selective colorimetric/fluorescent probe with aggregation induced emission characteristics for multiple targets of copper, zinc and cyanide ions sensing and its practical application in water and food samples, Sens. Actuators. B-Chem. 266 (2018) 730-741. https://doi.org/10.1016/j.snb.2018.03.089

[91]   L. Lan, T. Li, T. Wei, H. Pang, T. Sun, E. Wang, H. Liu, Q. Niu, Oligothiophene-based colorimetric and ratiometric fluorescence dual-channel cyanide chemosensor: sensing ability, TD-DFT calculations and its application as an efficient solid statesensor, Spectrochim. Acta. A. 193 (2018) 289–296. https://doi.org/10.1016/j.saa.2017.12.039

[92] K. Rajar, A. Balouch, M.I. Bhanger, M.T. Shah, T. Shaikh, S. Siddiqui, Succinic acid functionalized silver nanoparticles (Suc-Ag NPs) for colorimetric sensing of melamine, Appl. Surf. Sci. 435 (2018) 1080-1086. doi:10.1016/j.apsusc.2017.11.208.

Keyword Index

About the Editors

Dr. Inamuddin is working as Assistant Professor at the Department of Applied Chemistry, Aligarh Muslim University, Aligarh, India. He obtained Master of Science degree in Organic Chemistry from Chaudhary Charan Singh (CCS) University, Meerut, India, in 2002. He received his Master of Philosophy and Doctor of Philosophy degrees in Applied Chemistry from Aligarh Muslim University (AMU), India, in 2004 and 2007, respectively. He has extensive research experience in multidisciplinary fields of Analytical Chemistry, Materials Chemistry, and Electrochemistry and, more specifically, Renewable Energy and Environment. He has worked on different research projects as project fellow and senior research fellow funded by University Grants Commission (UGC), Government of India, and Council of Scientific and Industrial Research (CSIR), Government of India. He has received Fast Track Young Scientist Award from the Department of Science and Technology, India, to work in the area of bending actuators and artificial muscles. He has completed four major research projects sanctioned by University Grant Commission, Department of Science and Technology, Council of Scientific and Industrial Research, and Council of Science and Technology, India. He has published 176 research articles in international journals of repute and nineteen book chapters in knowledge-based book editions published by renowned international publishers. He has published 115 edited books with Springer (U.K.), Elsevier, Nova Science Publishers, Inc. (U.S.A.), CRC Press Taylor & Francis Asia Pacific, Trans Tech Publications Ltd. (Switzerland), IntechOpen Limited (U.K.), Wiley-Scrivener, (U.S.A.) and Materials Research Forum LLC (U.S.A). He is a member of various journals' editorial boards. He is also serving as Associate Editor for journals (Environmental Chemistry Letter, Applied Water Science and Euro-Mediterranean Journal for Environmental Integration, Springer-Nature), Frontiers Section Editor (Current Analytical Chemistry, Bentham Science Publishers), Editorial Board Member (Scientific Reports-Nature), Editor (Eurasian Journal of Analytical Chemistry), and Review Editor (Frontiers in Chemistry, Frontiers, U.K.) He is also guest-editing various special thematic special issues to the journals of Elsevier, Bentham Science Publishers, and John Wiley & Sons, Inc. He has attended as well as chaired sessions in various international and national conferences. He has worked as a Postdoctoral Fellow, leading a research team at the Creative Research Initiative Center for Bio-Artificial Muscle, Hanyang University, South Korea, in the field of renewable energy, especially biofuel cells. He has also worked as a Postdoctoral Fellow at the Center of Research Excellence in Renewable Energy, King Fahd University of Petroleum and Minerals, Saudi Arabia, in the field of polymer electrolyte membrane fuel cells and computational fluid dynamics of polymer electrolyte membrane fuel cells. He is a life member of the Journal of the Indian

246

Chemical Society. His research interest includes ion exchange materials, a sensor for heavy metal ions, biofuel cells, supercapacitors and bending actuators.

Dr. Tauseef Ahmad Rangreez is working as a postdoctoral fellow at National Institute of Technology, Srinagar, India. He completed his Ph.D in Applied Chemistry, from Aligarh Muslim University, Aligarh, India on the topic "Development of Nanostructure Organic-Inorganic Composite Materials based Sensors for Inorganic Pollutants". He worked as a Project Fellow under the UGC Funded Research Project entitled "Development of Nanostructured Conductive Organic Inorganic Composite Materials based sensors Functionalities for Organic and Inorganic Pollutants". He completed his Masters in Chemistry from Jamia Hamdard, New Delhi. He has published several research articles of international repute. He has edited various books with Springer and Materials Research Forum LLC, U.S.A. His research interest includes ion exchange chromatography, development of nanocomposite sensors for heavy metals and biosensors.

Dr. Mohd Imran Ahamed received his Ph.D degree on the topic "Synthesis and characterization of inorganic-organic composite heavy metals selective cation-exchangers and their analytical applications", from Aligarh Muslim University, Aligarh, India in 2019. He has published several research and review articles in the journals of international recognition. Springer (U.K.), Elsevier, CRC Press Taylor & Francis Asia Pacific and Materials Research Forum LLC (U.S.A). He has completed his B.Sc. (Hons) Chemistry from Aligarh Muslim University, Aligarh, India, and M.Sc. (Organic Chemistry) from Dr. Bhimrao Ambedkar University, Agra, India. He has co-edited more than 20 books with Springer (U.K.), Elsevier, CRC Press Taylor & Francis Asia Pacific and Materials Research Forum LLC (U.S.A) and Wiley-Scrivener, (U.S.A.). His research work includes ion-exchange chromatography, wastewater treatment, and analysis, bending actuator and electrospinning.

Dr. Rajender Boddula is currently working with Chinese Academy of Sciences-President's International Fellowship Initiative (CAS-PIFI) at National Center for Nanoscience and Technology (NCNST, Beijing). He obtained Master of Science in Organic Chemistry from Kakatiya University, Warangal, India, in 2008. He received his Doctor of Philosophy in Chemistry with the highest honours in 2014 for the work entitled "Synthesis and Characterization of Polyanilines for Supercapacitor and Catalytic Applications" at the CSIR-Indian Institute of Chemical Technology (CSIR-IICT) and Kakatiya University (India). Before joining National Center for Nanoscience and Technology (NCNST) as CAS-PIFI research fellow, China, worked as senior research associate and Postdoc at National Tsing-Hua University (NTHU, Taiwan) respectively in the fields of bio-fuel and CO_2 reduction applications. His academic honors include University Grants Commission National Fellowship and many merit scholarships, study-abroad fellowships from Australian Endeavour Research Fellowship, and CAS-PIFI. He has published many scientific articles in international peer-reviewed journals and has authored around twenty book chapters, and he is also serving as an editorial board member and a referee for reputed international peer-reviewed journals. He has published edited books with Springer (UK), Elsevier, Materials Research Forum LLC (USA), Wiley-Scrivener, (U.S.A.) and CRC Press Taylor & Francis group. His specialized areas of research are energy conversion and storage, which include sustainable nanomaterials, graphene, polymer composites, heterogeneous catalysis for organic transformations, environmental remediation technologies, photoelectrochemical water-splitting devices, biofuel cells, batteries and supercapacitors.

www.ingramcontent.com/pod-product-compliance
Lightning Source LLC
Chambersburg PA
CBHW061601220326
41597CB00053B/1560